MW00570426

CArlos KIBRIT

BREAKING THROUGH THE
PROJECT
FOG

JB JOSSEY-BASS™

BREAKING THROUGH THE PROJECT FOG

How Smart Organizations Achieve
Success by Creating, Selecting and
Executing On-Strategy Projects

JAMES NORRIE

John Wiley & Sons Canada, Ltd.

Library and Archives Canada Cataloguing in Publication Data

Norrie, James, 1965-
    Breaking through the project fog : how smart organizations achieve success by creating, selecting and executing on-strategy projects / James Norrie.

Includes bibliographical references and index.
ISBN 978-0-470-84071-9

1. Project management. 2. Strategic planning. I. Title.

HD69.P75N67 2008   658.4'04   C2008-900609-7

Production Credits
Cover design: Ian Koo and Jason Vandenberg
Interior text design: Natalia Burobina
Printer: Friesens

John Wiley & Sons Canada, Ltd.
6045 Freemont Blvd.
Mississauga, Ontario
L5R 4J3

Printed in Canada

1 2 3 4 5 FP 12 11 10 09 08

This book is dedicated to Paul, Lia and Jillian, whose gift of time and patient understanding allows me to indulge my passion for writing and sharing knowledge with others. I am grateful for your love and affection; my life is fuller because of our life together. And to Bev and Derek Walker (thesis supervisor), great colleagues and now good friends without whom this book would not exist.

Table of Contents

Preface

The writing of a book begins with an idea, and in this case, its origins belong to my many academic and consulting colleagues, incredible clients and the talented project management practitioners I have had the pleasure of working with over the course of two decades together. The field of project management is quite young; consequently, as "professional pioneers" we are all still figuring it out as we go!

This book is based on over 20,000 project hours; hundreds of online survey responses; and multitudes of case study interviews, training sessions and focus groups, conducted while completing my Doctorate in Project Management (achieved in 2006) under the able supervision of Dr. Derek Walker at RMIT. This book's genesis was my thesis for that degree. In addition, my consulting work in leadership and strategic project management within organizations continues as part of an ongoing research initiative at Ryerson University, where I am an Associate Professor in the Ted Rogers School of Management.

This book was written for the following audiences:

- *Executives and Board Members in the Private and Public Sectors*
  This group is ultimately responsible for strategy, governance, and an organization's overall performance. We know that the building blocks of executing any strategy take shape initially as projects, whether a new product or service development project, a new or improved business process project, or the installation of new IT or Web-related capabilities. We also know that organizations are investing millions, even billions, of dollars worldwide in projects and project management-related activities. For those tasked with selecting strategic projects or ensuring that investments in project management pay off for their organizations, this book provides practical help and insight into proven practices that work.

- *Project Managers and Program Leaders*
  Whether you are a sole practitioner, are part of a team inside an organization, or are at the top of your game and leading a global Project Management Office (PMO), all project managers need to continually challenge themselves to think about their ability to add value to their organizations. Within this book you will find a multitude of suggestions aimed at helping you become a strategic business partner to senior management, which is, in turn, deeply engaged in your organization's strategy and in the selection and execution of the strategic projects that really matter. You will also learn more about why some of what you have learned along the way may be limiting your professional contribution, success and career mobility. In short, you will learn how to clear up the project fog and excel as a respected professional contributor at the highest levels of your organization.

- *Consultants, Educators, Association Leaders*
  If you are responsible for shaping the future of project management and have an influence on this profession, this book will challenge you to think differently about the project management context and its place in business. You will also develop an understanding of what C-suite executives really want from their project management investments. This knowledge can help you develop and implement better project management practices globally that will ensure strategic results and executive satisfaction with your advice and contributions.

I have included many ideas directly contributed from individuals in exactly these roles—executives and practitioners from Canada, the United States, Europe, Australia and Asia, who shared their experience by anonymously answering online questionnaires.

I hope this book will free project managers to bring even more value to their organizations and move on to become part of senior management. When I talk with you at conferences, in training seminars, or while working with you in your organizations, I often hear your desire and willingness to contribute and your frustration at the lack of acknowledgement of the value of your potential contribution to strategy and its successful execution.

This lack of recognition must change, and it will over time. But it requires that we rethink our assumptions about strategic project management to ensure that they are relevant to the CEO and his or her senior executive team. The findings I share in this book and on the website (www.projectgurus.org) suggest a new mandate for project managers to participate at the highest decision-making levels of an organization. The website that the author and co-authors of this book have assembled provides free access for any project management practitioner or executive who wants extend the practice of strategic project management by joining a community of like-minded professionals.

We hope to change the current professional definitions of what constitutes the full scope of project management activities, especially in relation to strategic management of the project portfolio and project selection methods. This blurs traditional lines between strategy and project management and demands a fusion of two disciplines that is potentially disruptive, messy and ill-defined. However, navigating through the project fog is a worthy mission and one that will benefit both the individual project manager and the profession as a whole. This book and its accompanying website are hopeful first steps in that direction.

Meanwhile, the CEOs of large and small organizations in both the private and public sectors are frustrated too. They want project managers to contribute and they want their organizations to benefit from that contribution. But they are unsure of the essential value of project management and its role in the execution of strategy. This should concern everyone. For this book, I researched real behaviors of both management teams and project managers in an attempt to determine what *should* be happening as opposed to what *is* currently happening. I do not refer to this as "best practices" because the misuse of this term often confuses executives and leads them deeper into the fog. Rather, my mission is to help you find a path through the fog by providing practical, easy-to-use tools and techniques that work and are based on recognizing the reality of authentic organizational dynamics, politics and the "need for speed" that defines global competitiveness.

I hope that you can locate yourself in this book and see how it can contribute to the ultimate prize in business today—aligning your organization behind persistent effort that enables successful strategy execution!

For my academic colleagues: I am often dismayed by how little our insights into practice seem to matter to management. This should tell us something. Our deep understanding of the technology and techniques of project management is an asset, and our ability to influence practice, given that project management is an applied discipline that cuts across many traditional business subjects, is extremely relevant. Yet our insistence on the need for "statistical validity" and "quantitative proofs" means we risk not making as effective or complete a contribution to practice as we could. Ultimately, the study of project management happens best within an organizational context—action-based research techniques applying real-time interventions in real organizations that are geared to study differing results. While this method is gaining increased momentum and acceptance, it is not happening quickly enough. Project management is messy and complicated to study in real-world situations. I hope as academics that we can collectively work towards making our findings relevant in a business world where speed increasingly demands that we find newer, faster and more accessible ways of sharing our knowledge.

Acknowledgments

I wish to thank the many people without whose help and support this book would not exist.

First, I thank my collaborators for their contributions to this book. In one's professional life there are those that stand out for both their gifts and willingness to share them. The co-written chapters of this book (Chapters 5, 6 and 7) express wisdom that is the result of collaboration over many years with talented, dedicated and gifted colleagues with whom I truly enjoy working. You are each outstanding professionals in your own right and I am delighted to have worked with you on producing this book.

I wish to thank the many executives and practitioners from Canada, the United States, Europe, Australia and Asia who spent time answering questionnaires. Your honesty and willingness to share your experience were instrumental in the production of this book. Your willingness to participate in what has been, for me, a labor of love helped to clarify my thinking about strategic project management; I hope the final product justifies the investment of time you made in helping to produce the insights it contains.

I wish to thank my students at both the undergraduate and graduate level (and the students of my esteemed colleagues who may be using this book in the classroom). Teaching, more than any other activity, provides pure joy in my life. You challenge me not only to share what I know, but to justify the value of knowledge in practice. This gift ensures that I remain connected to my field so that what I do in the classroom remains relevant and engaging for you. I hope you will find this book is a launching point for your own careers in this wonderful and diverse field to which I have dedicated more than half of my life so far—afield that I hope will provide you with a diverse and satisfying career as you move from project management neophyte to leader in the coming decades!

To the wonderful folks at Wiley Canada, and an especially gifted and understanding editor (I mean you, Karen!), my thanks for making this project a reality and for enabling me to collaborate yet again with Michelle Nanjad, who is as talented and patient a taskmaster as anyone I have ever worked with. Your collective influence is evident in this book, and so the credit for success is duly shared.

To my readers who approach this book in the hope of becoming better at what you do, my sincere thanks and appreciation. I would like to hear from you: please visit my website and send me your comments, questions, and concerns at www.projectgurus.org.

Toronto, Canada
November 2007

## CHAPTER ONE

Spotting Project Fog

The last five years of my professional life involved completing a doctorate in project management (DPM) and exploring strategic issues related to the practice of project management in organizations. The research conducted for my thesis is the genesis of this book. As I traveled the world of project management and C-suite executives that are so critical to the process of sponsoring and executing projects, I began to recognize that both parties were frustrated with the efforts and results from the other.

Project management professionals often feel distant and unconnected to the strategic management processes of their organizations. Rather than being strategic partners, they feel like a pair of hands that kick into action once a project has been selected and approved for implementation. Yet they could contribute so much more to the process if they were involved from the outset in the definition and selection of projects and worked closely with the executive team to help them realize their strategic vision. Since projects are often the building blocks of strategy execution, something that all executives should be focused on, it would seem like a natural partnership. Project managers want to ensure that their hard work matters and that what they are doing is truly strategic. Yet in my surveys, many project managers worldwide report feeling like they work on a never-ending list of projects with doubtful intentions and lack a clear strategic focus. They rarely feel connected to their executive sponsors or feel that they are on a strategic mission together towards a defined destination. This feeling of drift is not healthy for any professional. Its origins need to be examined so that organizations can address their level of motivation and engagement.

Over the past twenty years there has been an unprecedented level of investment in project management, as more of the work of most organizations comes in the form of projects. Accordingly, executives have stepped up and approved investments in new resources in a wide variety of ways.

This can mean funding additional dedicated PM positions or functions, such as a project management office or center of excellence. Some organizations buy, and some build, standardized project management methodologies to support consistent project management practices internally. Others have invested in online tools and other forms of automation to support PM processes in the belief that project management practices, properly implemented and executed, improve productivity and returns. However, the reality for many CEOs that I talk to is that they are not always seeing these returns clearly. In fact, there are almost as many published studies about how many strategic projects fail at huge cost to their organizations as there are articles about the success of project management practices in organizations. So all this leads to valid concerns about the value of all these PM-related investments and just exactly what the benefits are. What is really going on? I believe the answer lies in the presence in most organizations of *project fog*.

Project fog happens when we have so many things we call projects (or initiatives, or programs), picked from an even larger list of potential concepts, approved or not over the space of many years, and now overlapping to the point where the many potential business benefits may or may not still be valuable. We lose track. We get lost. Like real fog, project fog creeps up on us and overtakes us almost unaware and before we know it, it's too late to avoid it because we're in it.

Oddly, most of the organizations that I studied do not actually have a "project management problem" per se, even when they report feeling like they are in a project fog. In fact, most organizations have become pretty good at executing single and multiple projects; this is not where the problem lies. So my research focused on identifying the real underlying problem and trying to understand why executives and project managers feel the way they do.

As I determined from various case studies, the origins of project fog occur well before an organization begins to execute projects, and instead relates to how projects are proposed and selected. In project management terms, this is referred to as project portfolio management (PPM)—an emerging buzzword in the profession today. However, while there is substantial discussion about this important aspect of project management, little that is definitive has been written about how to do it well, and this is both the source of and solution to the project fog. If we learn to manage the ways we propose and select projects as well as we know how to execute them, the project fog will clear up and your organization's strategic potential will improve. So let's begin.

HOW THE FOG ROLLS IN

Spotting a portfolio management process problem among my clients often begins with a simple complaint: "We have too many projects and project proposals!" (This is the salient symptom of being in the fog for most organizations.) How is it, I wondered, that these organizations were selecting and then approving so many of the same strategic projects they were now complaining about in the first place?

As I discovered during research interviews, there were in effect so many forms of effort inside most organizations called a project—a program, initiative, pilot, campaign, product launch or a host of other names—that losing track of them was easy! The senior executive team obviously must think they are all "strategic" or they wouldn't be approving them, and project managers must have been consulted on the plans for them. So then who exactly is responsible for an organization having "too many projects"? Those actually proposing them or those ultimately approving them? The answer, it turns out, is both. I began to refer to this in my consulting work as "project fog," and the term seemed to really resonate.

This also led to three simple questions that I pose to organizations I work with to help them determine if they might be in the fog. These questions are:

1. Are you planning the most strategic projects possible, and how do you know that as an executive team member or project manager?
2. What if the most strategic projects are not being conceived and proposed because your project selection criteria mostly encourage projects that are financially efficient versus strategically effective?
3. And what do you do if your organization's primary purpose cannot be measured in terms of profitability? How do you measure strategic contribution and connect this to project selection?

Using these questions, I began to notice that the portfolio project management approaches that organizations were using to try and manage all of this effort seemed inadequate to address the true complexities of enterprise-wide project selection and management. In fact, if they had a systematic approach to project selection at all, it seemed to be one that validated simply having lots of projects!

Too many projects were getting approved, and for all the wrong reasons. And then they were not managed cohesively for maximum strategic benefit. All approved projects were chasing the same limited resources available to do *all* the project work, so some were bound to fail. Project chaos was exhausting everyone. It became clear to me that new strategic project portfolio management practices were needed.

I realized that, for all of the effort to professionalize project management, maybe we had missed a strategic underpinning of huge potential value: what if we were actually expending all this effort to execute projects that were potentially not very strategic? And what did this mean for the profession? Could avoidance of defining solutions to this problem be a lapse in professional responsibilities? Similarly, as executives or board members, what could you contribute to solving this problem and encouraging a level of partnership with project management professionals that would focus on strategy execution rather than purely on project execution?

To clarify the extent of the problem, my initial research efforts focused on a few large corporations. What became immediately obvious was that their business objectives were normally related to enhancing profits. In and of itself, this is not a surprising objective, nor would most shareholders say this is inappropriate for a private sector firm. But when I looked more deeply into this challenge, it became instantly clear that if an organization's effort at project scoring, a critical aspect of picking and prioritizing projects, was based exclusively on financial efficiency, it was then sub-optimizing strategic outcomes. Why? Because the most strategic projects were unlikely to always generate the highest return. Ask any CEO and he or she will tell you honestly that innovation and creativity cost money—and that is exactly what strategic projects are all about. They are vital to long-term health but may actually cost us in the short term. (So while you can probably agree with this statement, I expect that your project selection system today likely defeats this outcome by focusing on the selection of projects with high financial returns.)

This was intriguing on another front, because almost all of the organizations I studied had complex, written strategic plans, and a review of most of these would reveal fairly complex underlying business acumen. Yet few of them had non-financial measures as part of their project selection criteria, as seen in the graph below.

Planning & Measuring Paradigms

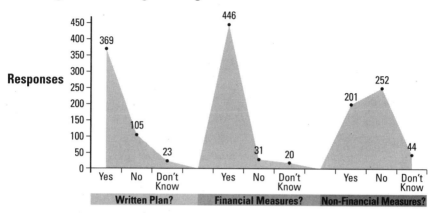

The shape of this graph is important, for it gives visual representation to data that underlie an important trend. The shape of the first area of the graph (on the left side) shows 74% of respondents (369 out of 497) had written plans and the second part of the graph (moving towards the right) indicates a whopping 90% had financial measures associated with those same plans. Yet we see quite a different story when it comes to the use of non-financial measures in the last graph. Here there are both more diffuse levels of usage and a much higher number who simply responded that they "didn't know." This began to suggest that the measures being used to track strategic success were perhaps less robust than the underlying strategic plans they were attempting to support. As we will see in the later case studies, this was exactly the case.

So the consequence of this relatively naïve approach to project selection is that if an organization uses only financially based metrics to pick their projects (the case for the majority of our survey respondents), this can lead to a false belief that they are making the right decisions and selecting the right projects—when the opposite is probably true. As you will see, in the long term these organizations would be losing opportunity (and the associated profits!) when you consider the results across their entire project portfolio. Was it possible this was happening because they were unaware of this loss, or were they simply unwilling to do the harder work of figuring out what was truly strategic?

Another perplexing finding was that almost all of the existing PPM methods that I saw were focused on how to rank and select new projects, yet they paid little attention to shifting the priorities of (or canceling) existing projects. This gives rise to a "portfolio within a portfolio" approach, where the project selection method does not operate on the entire portfolio at one time, but on a sub-set. Regardless of what sub-set we define (buckets by project type; separating new projects from existing ones; or using definitions of strategic complexity, such as platform versus operational projects), the point of managing the entire portfolio as one is lost. An additional layer of financial and resource inefficiency has now entered the picture. Just eliminating some of the redundant work done on projects that should be canceled in favor of new initiatives can significantly improve project costs and resource utilization. Yet few seem to be concerned about this. Most project managers I spoke with during my research reported that organizational priorities were often on new projects rather than on the status or progress of existing ones.

When this same practice of false economies is translated into the more complex strategy-making context of a public sector or not-for-profit organization, it becomes even more inappropriate as a project selection method. In fact, it might even be considered criminal or at least unethical as a recommended practice in that context. The problem arises from an underlying assumption that projects which maximize financial return and minimize risk are the most strategic projects to select. This premise can have unintended or even disastrous consequences and numerous examples could be cited in most countries of the world of this very phenomenon. For both types of organizations, little seems to have been written to date that can help guide the definition and selection of "strategic projects," making the application of portfolio management tools hard to achieve in practice for just about any organization of any type.

Now let's thicken the fog slightly: when I directly explored the current use of PPM techniques among experienced project managers and executives, there was a decidedly negative response. Most either didn't know too much about it or reported generally lower use of this technique than project/program management and project management offices (PMOs). There was a sense, later confirmed in detailed interviews, that most of the approaches in use today were seen as too complex for the limited benefits they seemed to deliver once implemented. This hypothesis was further confirmed by an early pilot study conducted among experienced PMs who clearly shared this view.

So this is not simply a matter of lack of experience with methodologies or techniques that some were not aware of; rather, it now seems like a genuine gap in professional practice. This means that without substantial improvements, most organizations will not choose to use portfolio management techniques as they exist today. And those hardy few who do risk a failed implementation. They only frustrate their organizations' executives, project managers, and project staff. Not such a good outcome, for a profession supposedly on the rise. So . . .

The benefit of solving this problem would be high for organizations, and would be immediately evident to most executives—who would welcome the resulting clear focus on strategy execution. The trick that remains is to solve the problem and help organizations pick the most strategic projects for execution.

THE FOG THICKENS

This book is based on an extensive multi-year study of project managers and executive sponsors. The executives were working in a range of organizations, varying in scope and complexity, including financial services, insurance, information technology, transportation, manufacturing and industrial services. There were also respondents from government and the not-for-profit sectors. Given the hyper-competitive nature of the global economy today, all of these executives reported an urgent need to exponentially enhance organizational performance by improving the alignment between senior executives' strategic goals and project managers' efforts. Nobody has any time to waste and everyone is trying to move ever more quickly. To make this work requires a partnership around co-execution of strategy. It takes integration and mutual support.

On the other hand, project managers are worried about becoming mired in the "strategic muck" and unable to escape involvement in aspects of organizational strategy that they may or may not understand, and for which they do not feel responsible. They report that executives say they want to be strategic, but their behavior sometimes indicates otherwise. Some might actually prefer to operate in a strategically ambiguous place, where nobody can actually determine whether or not they really are as smart and accomplished as they seem. The problem demands action from both sides.

So how can we strengthen the link between project selection methods (primarily the domain of project management professionals) and an organization's strategy (primarily the domain of executive teams) so together we can improve results?

When we only do the projects that really matter, internal efforts to improve project or program management will also matter. This becomes another important part of the project fog experienced in many organizations: they increase their spending to implement project management practices, but this does not necessarily translate to improved performance. Project managers are not yet fully engaged strategically within their organizations. Until we are willing to step forward and help our organizations pick the right projects, no amount of effort on our part will matter.

It is also clear that nobody is creating the project fog deliberately. There was no simple, single cause I discovered in my research that will solve this dilemma for organizations. In fact, just the opposite was true. The problem is complex and as such a uniform solution applicable to every single organization is unlikely.

As with actual fog, project fog creeps up on organizations slowly and by the time we see it, it's too late. Before we know it, we cannot move forward at the speed we want—until it clears. Think of the never-ending series of project approvals, new project proposals, a constantly limited resource pool to draw on to complete them—and yet the shared determination to believe that all projects are equally important never falters. We don't stop anything. Bad projects never fail—they are simply absorbed, renamed or morphed into new projects, or into "strategic programs." We toil on, hoping for success from a sufficient number of projects or programs, hoping that the organization will somehow accomplish its strategy. Does this sound familiar? Does this sound like the kind of deliberate strategic approach you want to have, as either an executive or project manager?

While various parties, including the Project Management Institute (PMI), have tried to address this reality through suggestions on practice (such as their recently released practice standard for PPM in 2006), it is unlikely that the fog can be wisped away purely on the basis of standardized practices. The fog is actually as thick as it is because the problem involves a combination of people, process and technology. What is required is that we adapt behavior to orient it towards mutual support and partnership, that we have a defined process with a framework for finding organization-specific solutions rather than generic one-size-fits all prescriptions and that, over time, we find technology solutions to support all these measures. This book is a first step on that journey and I hope it will help you and your organization find out how PPM tactics can improve performance in your own organization.

THE RESEARCH PARTICIPANTS

In addition to the individual case studies we will explore in later chapters, I conducted a significant online survey over two years that garnered responses from 497 executives or practitioners from organizations both large and small around the world. Some of these respondents also accepted follow-up invitations to participate in phone interviews or panels that in turn form the basis for much of what is reported here. Since so many of the insights I am going to discuss come from this participant group, let's take a quick glance at its make-up:

Respondents' Title/Level

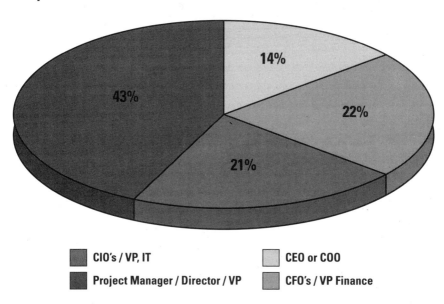

| CIO's / VP, IT | CEO or COO |
| Project Manager / Director / VP | CFO's / VP Finance |

You will see that the majority of the respondents are actually non-project managers (57%), spread across three distinct titles, with the remainder of the respondents (43%) involved in project management roles. This makes the group quite balanced and likely able to provide insight from both the executive and project management perspectives on the challenges they face in their organizations with regard to PPM practices.

And it would appear that, regardless of who you speak to in the C-suite of most organizations (CEO, COO, CFO or CIO), they report some discomfort with their project management results, particularly the link between strategy and project management. You will find additional insight into this problem in the chapters that follow.

Number of Employees

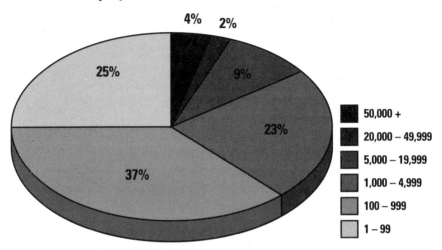

Project fog in some form also appears to have no particular boundaries in terms of the size of organizations it affects. Aspects of the problem were reported in organizations as small as 100 or fewer employees, all the way up to global multinationals with more than 50,000 employees. The challenges are universal in nature. This contrasts with the views of other researchers in this field, who have suggested that PPM challenges vary both in their value and presence as a technique in companies of differing sizes and industries. In my view, the scope and complexity of the solution needed to address the larger scale of big organizations varies, but the principles of how to address the problem remain the same: regardless of the organization's size, the features of the problem are the same.

Finally, the different kinds of business models we find in organizations are an important consideration: are government and non-profit organizations experiencing the same kinds of issues with strategic project selection? Do their perspectives differ significantly, and, if so, why? I have made a deliberate effort to elicit valid responses from public sector executives. The sample for this research includes a broad cross-section of organizations in the public and private sectors, resulting in almost an even split among respondents as seen on the associated graph.

Since the problem is reported consistently among respondents and, since the survey sample is quite demographically representative, it leads me to conclude that the problem is likely present in all sectors to some degree. The wholesale adoption of "projectization" as a method

Organization Type

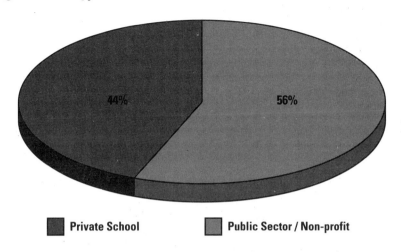

Private School   **Public Sector / Non-profit**

of structuring organizations and the work they do has led to a virtual explosion of interest in and application of project management practices around the world. But does this necessarily mean they are being applied in a strategic way?

There has never been more demand for professionals with our skills; this is a good thing for our future development. The continued lure of the profession has led to explosive growth in membership within the primary professional certifying bodies, such as the PMI, in the last few years. We've seen this also with the International Project Management Association (IPMA), with the explosion of companies offering project management training and technologies, and with the number of people who either describe themselves or are formally described as full- or part-time project managers in organizations. Still, the very real problem of project fog exists and will continue to grow unless we invent new ways of working together as project managers and executives to solve this dilemma of good intentions but bad execution of too many projects. We need a universally applicable PPM process that can become part of the project management body of knowledge that provides for a common framework while recognizing that the implementation of PPM in each organizational context will be different and unique.

DEFINING A NEW STARTING POINT

An organization's environment, its own choices about how to structure itself to accomplish its strategy, its scope and its scale all make the starting point for PPM in any organization unique. No two organizations are exactly the same. This means that PPM, as a strategic process, will likely not be prescriptively the same for every organization. For a profession like ours that so often tends to define itself by fixed practices and "one size fits all" solutions, the notion of a methodology that could accommodate each organization's particular needs is daunting. In fact, some professional associations and societies may be scared away from even attempting to include such diversity in their definitions of project management practices for this very reason—but that is a mistake.

As my study progressed, a general process framework emerged that helped me consider the many distinct aspects of the PPM *as they should be* in practice, rather than *as they are* found in practice today. The dimensions of the framework shown below can easily be drawn as a process diagram. The model works well in practice, while remaining theoretically sound, with respect to the origins of modern portfolio theory.

Throughout this book, we will explore each aspect of this framework in greater detail and explore the practices, methods and specific tools that

A Better PPM Model

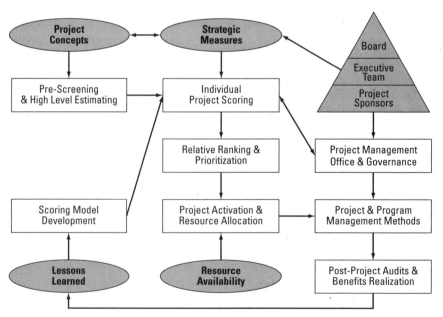

complement the model in practice. My goal is to enable you to use it in your organization to make a strategic difference Let's consider it from a high-level perspective first.

Circles represent inputs to the process. Boxes represent process steps or methods that are invoked in relation to these inputs. The triangle represents the major stakeholders and the flow of their interaction with various inputs and process steps. For most readers, I hope the diagram is self-explanatory. What is not clear yet, but will be in time as you continue reading, is the complexity of the process steps themselves and what is required to support and complete each step.

Overall, the three phases of the framework essentially involve

- the initial conceptualization and project planning phase;
- the prioritization and activation of approved projects;
- the tracking, reporting and auditing of project results.

In and of themselves, these tasks are not new. What is new is the presentation and order suggested by the framework and suggestions from the research as to what constitutes effective and efficient practice for each step. Throughout the rest of the book, I will make frequent reference to this generic PPM process map as a guide to our explorations of how to make these concepts work in practice.

Because of the complexity of organizational types and of the activities underlying the process steps, defining an effective, flexible, yet comprehensive PPM solution is a challenge. Yet it is the only way forward for us as a profession. There is too much ambiguity today among practitioners globally about what actually constitutes portfolio management. So to make progress, the scope of portfolio management needs to be clear to those who depend on us for this advice—C-suite executives. I have found *three specific benefits* that occur when you successfully optimize three inter-related tasks: originating, selecting and prioritizing projects, as suggested by the process steps of the framework. I frequently use these three benefit statements in explaining the value of this approach to PPM when working with CEOs. They are:

- Improve the ways projects are conceived and submitted for early approval to avoid wasting effort on non-strategic projects.
- Define practical methods to directly link project outcomes to a balanced set of measures that optimizes and aligns the project portfolio with your strategy.
- Define how to govern and manage the project portfolio with respect to clear project prioritization, staged activation, tracking and reporting to realize maximum benefit, control costs and reduce risk.

Any C-level executive will tell you that any business process has to be simple and useful to actually work in practice, while remaining theoretically sound and delivering on its promises. The failure of so many "perfect processes" is that their costs outweigh their benefits and they are abandoned or never fully implemented as a result. Conversely, when dealing with a process as complex in practice as PPM is for most organizations, you cannot oversimplify the process steps or you risk a non-strategic outcome because the process is incomplete. You should seek in any workable process to design a balance of the efficient and the effective within the realm of cost and reliability that is affordable for your organization.

This challenge is illustrated in a simple but powerful diagram that I often use to explain this issue to my clients. The two competing forces to contend with in the design of a PPM process is the balance between tactical and strategic outcomes and between the simplicity and complexity of the process itself. This is shown in the diagram below.

The Range of Current PPM Methods

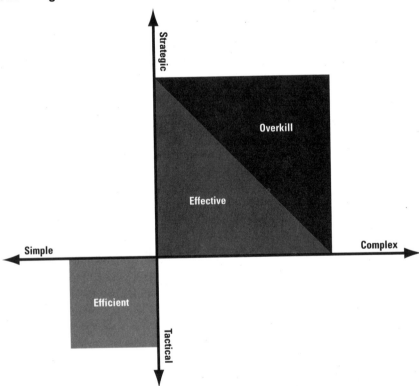

After examining the various methodologies and software tools currently available in the global marketplace, I was able to determine that if the portfolio management solution presented is too efficient, it will yield a result that is too tactical and thus ultimately unable to support the organization's strategy effectively. This trade-off is not worth the apparent simplicity the method offers. This is shown in the lighter area of the diagram on the bottom left-hand side.

PPM methods that rely too heavily on selecting projects only on their financial returns, for example, fall into this zone. While these pseudo-scientific solutions may seem seductive because of their precision, they provide a false sense of security for executives who rely on them and are often based on false premises that can impair long-term results. Solutions and methods like these are the most dangerous of the PPM solutions in use today because they distort and diminish the value of strategic project selection and reduce it to a kind of simple "math-a-magic," often supported by software vendors who want to sell you software rather than help you solve your project selection dilemmas.

Conversely, solutions that delve by design into strategy in minute detail in an attempt to achieve perfect strategic alignment are also not good solutions. By frequently including all proposed projects in their scope rather than focusing only on larger or more strategic and costly initiatives, they are likely to be seen as unwieldy, unworkable and generally as overkill by most executives. For instance, projects that are mandated by law (e.g., updating employee deduction tables on a payroll system to account for annual changes in limits and deductions), while they meet our definition of a "project," are neither truly strategic nor subject to choice about when and how they are done.

This is where our bias for perfection of process as a discipline can lead us astray: the pursuit of the perfect process cannot be justified in time or cost, except for organizations where there are major consequences (ecological disasters, death or destruction, for example) that make a perfect process a worthwhile pursuit. While detailed processes like this claim the benefit of bringing structure to complex tasks, the benefits of processes like this are not worth the effort to most organizations and should generally be avoided.

So what do you need to do to succeed? The answer can be found on the frontier between the two competing axes: a balanced solution between efficient and effective that enables the strategic outcomes of the organization, while remaining simple enough to be practical and workable. Indeed, and as the research confirms, any potential PPM solution in most organizations would have to be acceptable to both project managers *and* senior executives if they are expected to work co-operatively to implement and use them.

An organization's strategy is not created in a vacuum. There are the normal inputs of an organization's planning framework and the obvious input of its competitive position as natural considerations in the strategy-making process. However, in most organizations there are other (often unvoiced) factors that also play a significant role. These include:

- Any applicable policy frameworks an organization may be subject to (particularly in regulated, monopoly, or government organizations);
- The politics within the organization and the degree of executive or leadership self-interest they are willing to impose on the process;
- How the organization procures its vendor/partner relationships.

In the public sector, there are often more formal procurement practices; in more entrepreneurial organizations, the selection of a partner or vendor may actually be a factor in how the ultimate strategy is set, because of the informal or cozy nature of these relationships. Regardless, as a professional in an organization, you need to be aware of these kinds of strategic nuances within your organization.

So, clearly, you cannot select projects for strategic impact unless you take into account all aspects of the strategy-making process of your organization and its competitive context, again reinforcing just how specific rather than generic any PPM process you design must be, if it is to be effective in helping your organization achieve its strategic goals.

THE LINK BETWEEN CURRENT PRACTICES AND "BETTER PRACTICES"

The term "best practice" has always struck me as arrogant and unrealistic. The more appropriate term is "better." In any profession, practices should constantly evolve. Practitioners who are dedicated and creative are always pushing for new ways to make their work better. But this also suggests an ongoing process. One of my clients uses the phrase "transformational practices" and I also like this. As new knowledge permeates the organization, it transforms the way we work, because we realize that there is a better way.

The corollary of this is common practice, if learned on the job, may not necessarily be the "better practice." In fact, without continually evolving your knowledge of current practices, relying only on what you already know might increase the gap between your current practices and better practices. This is true for much of what passes for sound PPM methodology today: it's moving so fast that methods that looked great a few years ago can be significantly out-of-step with what more successful organizations are doing now to outperform their competition.

That is not to say the practice that is most common today will not work; only that a better practice may exist. As an academic in the field of project management, I find there is a risk that research undertaken at either extreme of this spectrum would either be too simplistic to make a difference in practice (i.e., it is theoretically perfect but of limited real value) or too complicated to inspire its use (theoretically possible but impractical in the real world). Neither outcome is desirable.

So how do you benchmark where you're at in comparison to better practices? And is there a better practice in use today around project portfolio management that you need to learn about that could be of value to your organization? Obviously both of these questions are essential.

So I began to explore what had already been done. This is easy enough in a field like project management, where the body of literature is not so vast; one can easily search for information about perceived "best practices." Rather than new practices displacing the old, I simply build on a continually expanding base of knowledge to incorporate new findings about what works well. The best organizations around the world continuously improve and share knowledge and I believe the same principles can be employed by project managers.

When it comes to project portfolio management, the majority of what has been written about its implementation has tended to focus almost exclusively on relative project-to-project comparisons. This approach proposes that new projects be compared to each other and the "best ones" somehow picked. Or perhaps projects are put into "buckets" and assessed relative to their peer projects, and the "best ones" float to the top of the to-do list for that bucket. Of course, a multitude of models have been proposed to date, but most have the inherent flaw of only being able to pick from the projects proposed. *But what if the projects being proposed are insufficient to accomplish your strategy? Do you not risk relying on a portfolio that is not robust enough to accomplish your strategic goals?*

While any relative comparison methods are at least a starting point, they do not actually help a company select and manage a truly strategic project portfolio that ensures the successful execution of its overall strategy. All these methods can do is ensure that you create a relatively higher performing portfolio of projects from what is being proposed. But what if the most strategic projects are not being proposed, because your project-selection criteria don't encourage their conception in the first place? This potentially fatal flaw in logic—not connecting project scoring and selection to an organization's strategic intent—should cause you to be suspicious about some of the PPM practices being recommended today. Now that we know what is at stake, the important question is *what should we do about it?*

One thing we need to do right away is move away from purely considering either current practice or academic theory and instead fuse them together. This unification of theory and practice will yield solutions that are sound and also useful and practical to implement. So my objective with you is clear: my research proposes a valid theoretical strategic scoring model that will work to enhance both the clarity of PPM for most firms and also improve the probability that the portfolio of projects ultimately selected by the executive team will optimize execution of their strategy. This breakthrough will allow organizations in both the private and public sectors to improve strategy execution and results. But to do so, we need to start with a clear end in mind—and this means linking projects to measurable strategy.

## CHAPTER TWO

The Importance of Being Strategic

I do not see an effective PPM process as including any scope beyond the framework set out in Chapter 1. However, this narrowly defined generic process framework can make a spectacular contribution to strategy execution—something many executives struggle with in today's hyper-competitive business environment. Project managers who contribute directly to accomplishing this objective can get executive recognition for their ability to navigate through the project fog and reduce all this confusion about what projects really matter.

Obviously, any valid and reliable approach to PPM needs to ensure the most strategic projects are being selected and prioritized for execution. This implies a centrality around the clarity of strategy in the first place—something that is not always present—and a degree of consistency in the strategy that enables others to be aligned to support it. Easier to say than to do!

IS YOUR ORGANIZATION ALIGNED FOR RESULTS?

It is important to know where you are going, as well as what method you will use to determine if you have arrived. You need a "strategy map" to figure out where your organization is going.

In any setting, this must start with an understanding of your organization's long-term mission, its current strategy, its performance goals around all key processes and then, using a gap analysis, go on to determine what it will take to get your organization where it wants to be. Choosing the projects and programs that will get you there is called *strategic alignment* and is a key concept borrowed from general management practices,

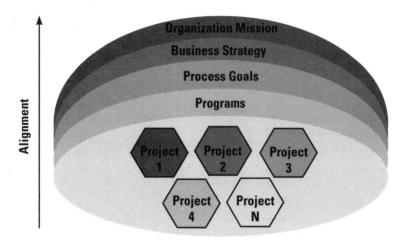

translated into project management terms. A simple diagram related to alignment as shown above might help.

During the research for this book, it became clear that alignment between the projects chosen to proceed and an organization's strategy is of paramount importance. Yet most organizations get sub-optimal results. For instance, one of my clients had planned and implemented a variety of technology-oriented projects for their call center but was frustrated about why there was no apparent improvement in key indicators related to the call center's performance. When they aligned key processes to key indicators and tied these to compensation of agents, remarkable improvements occurred. In retrospect, this was a perfect example of this phenomenon—until the projects and programs associated with the call center were aligned to the core processes and strategy, nothing of note was going to occur.

CURRENT PROJECT SELECTION METHODS

Particularly in the private sector case studies, I saw what appeared to be an important phenomenon: in the absence of specific scoring methods that allowed them to measure the contribution of specific projects against intended strategy, the executive teams adopted easier, but non-strategic, criteria to select their approved projects.

In fact, the survey research shows that current methods of picking projects in most organizations rely on a relatively narrow set of similar criteria, shown in the following chart. This may explain some of the confusion and fog.

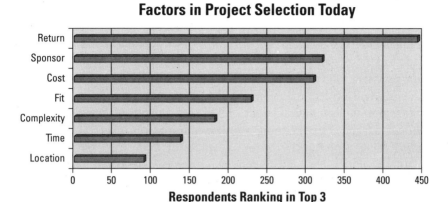

Factors in Project Selection Today

The chart tells a story similar to what you may already suspect: financial returns, measured on an individual project or program basis, are the primary project selection criteria in use today. But even more interesting are the other factors that are in use.

The most obviously non-strategic example of a criterion that would not likely be associated with picking strategic projects is to base the selection on who proposed the project! Yet this is the second most cited method. This is not to suggest that most executives are not competent and committed; but it would be fantasy to suggest that, merely on the basis of personal judgment, a project should rise to the top of a firm's priority list. And we know something else: given that many executives travel within similar circles, we often see projects arising as a result of "copycat syndrome," where an executive becomes aware that a competitor or collaborator is doing something and suddenly feels a need to explore its application within his or her own organization. This may result in a false sense of security that we are "keeping up" strategically.

Similar flaws in logic are easily found for any of the other factors cited on this list by survey respondents. It is clear that the extent to which this is happening in organizations must be causing some of the project fog we are all experiencing today! Yet this cycle can be hard to break.

While we can conclude that many organizations do not have any kind of consistent portfolio management system in place today and could benefit if they implemented a more strategic project selection framework (instead of using their personal preferences for project decision making), this also implies accountability—not only for the process of project selection, prioritization and activation, but for the original strategy and its outcomes. Not all executive teams will embrace this responsibility, and this makes the potential contribution of project managers a strategic matter for the organization. In fact, partnering with the CEO or other senior executives to make this happen can be a defining career move for bold, confident project managers.

The public sector case studies touch on a common problem found in this arena: the overwhelming number of initiatives or projects that are planned for completion by central agencies or bodies in comparison to available capacity in the field. Given the higher level of resource constraints in the public sector, it is not a surprise that this is so clearly a challenge. Again, the adoption of PPM helps address this problem, but perhaps the benefits in this area have been underachieved because the scoring models were relative instead of absolute. It is much harder to stop or deny a project simply on the basis that it offers a relatively lower return than others. The only way to do this is to fall back on a resource-constrained view and simply not do the projects which fall lower on this prioritized list, an outcome that is endorsed by many of the PPM methods suggested in the professional literature today.

This creates a massive incentive for project sponsors to attempt to manipulate either the scoring of their individual project or the cut-off of available resources to ensure their projects get selected and approved. The case studies demonstrated that this was exactly what was happening in practice, and the same project resources were being forecast for too many projects. Resources were stretched thin; projects were failing.

THE STRATEGIC IMPLICATIONS OF LIMITED RESOURCES

If your PPM process treats capacity as a fixed input rather than as a cumulative asset, it means capacity is ultimately defining your organizational strategy. Again, this is a seductively simple but sub-optimal method that abandons true strategic trade-off decisions. In my consulting practice, I call this "pseudo-strategy," because it looks and feels strategic but ultimately is not.

If you limit projects and then select and approve on the basis of project budget, rate of return, or the use of scarce resources, you will complete projects that consume current resources more efficiently. However, this selection process will be distorted because higher-value projects that may ultimately be more strategic and effective for the organization to complete will go unchosen. At the project level, this is similar to the classic

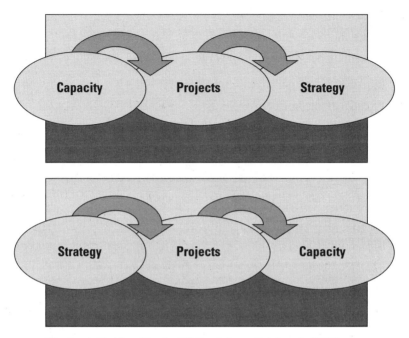

Strategy-Making Trade-Off Decisions Arising in PPM

"efficient versus effective" trade-offs we see in other areas of business. To be strategically effective, we must let go of the notion of capacity as a strategy-defining criterion and instead use measures associated with our strategy to pick the optimal portfolio and then determine how to fund and execute it.

The opposing positions of strategy and capacity tell the entire story: one must be the driver and one the outcome. The design of your organization's PPM process and how resource considerations fit into your process design are important determinants of how strategic the outcomes of the process are likely to be.

This brings us to another frequent practice: putting the organization into a "scrum" to argue or advocate for access to resources before selecting the "winning projects." (Or, depending on your corporate culture, these could be labeled the "whining projects," since they often get picked because of all the whining done about their importance—without any substantiated facts!) This process is non-productive and highly political in nature, and it won't yield an optimal selection result. It's more like a run-off voting process: eventually you get to the lowest common denominator that is acceptable to the majority. In fact, when we combine this approach with the fact that the influence and power of the sponsor is a factor in many organizations in situations like this, we end up with a few key executives getting their way and setting the subsequent strategic agenda. Again, many of the interviews we conducted during the case studies confirmed that prior to the implementation of the new PPM framework, this was how many organizations approached project selection.

This approach is especially common in sales and marketing. In many of our case studies, arguing against these projects led to claims that there was no other way to meet revenue or sales targets. Yet, once PPM was installed and these same projects were not approved, in most cases revenues and sales either stayed predictably flat or increased! This is a powerful example of why we need to base approval of projects on more than intuition or threat.

When we use a method that demands a neutral justification of the project's merits against pre-established criteria of strategic contribution, it is more likely that the internal conversation will shift to the more positive focus of making sure that all the strategic opportunities available to the organization can be accomplished. While this demands that we be ready to make our strategy measurable, it is worth it.

Voices of Experience: Vice President of a Major Bank

Bob is the vice president of a major financial institution in the U.S. and responsible for an annual project portfolio in excess of $100 million. We worked with him and his team from early 2000 until 2002 on implementing PPM at the enterprise level.

What do you see as the role of project management in terms of your business?

While I did not begin my career in project management, who would have ever believed I would end up here? I was surprised by how passive project managers are about the lack of specifics they often get from senior managers. Half the time it's not at all clear what we are trying to do. So we have to demand that they clarify what they are trying to accomplish and then simply use us as tools to get this work done. We are really good at the tactics—they have to be just as good at the strategy. This means that while we may not be directly involved all the time, even if we should be, we have to intimately understand the business goals and objectives of our project sponsors if we are to support them strategically. This probably means telling them the truth about selecting good projects—they don't always get it and we can help them do it better if they get us involved more. But we have to convince them this will be valuable and not just an exercise in making things too complicated, which is often the case with project management.

How do you pick the project managers you work with?

First of all, I do not believe that project management is a completely generic skill. In fact, without some business knowledge of financial services, I fail to see how a project manager could actually do a good job around here. The business would simply push them around and take advantage of the fact they don't know anything about banking. So I put a premium on picking professionals who have both excellent project management skills and excellent strategic knowledge of our project content. This works well. I also pick project managers who demonstrate an understanding of the project management process as focused on delivering outputs rather than as a series of inputs. Good project management is not about a bunch of checklists and forms. It's about getting complicated work done fast and right. And it's not an easy career...so to really answer your question, I pick project managers who aren't really project managers. They are simply good managers who happen to pick to work on projects!

What's the biggest challenge we face today as project managers?

I stumbled into this field but would likely have not have chosen it. There are days that it's pretty strange for me to recall what we did before we had project management, but more and more we are structuring work and jobs at the bank as projects, creating demand for people with good project management skills to lead them. This means we are going one of two ways in the future: either to a place where every manager will simply have the fundamental skills necessary to manage a project and we will dispense with specific resources assigned to this task and simply have more managers. This is a little bit like going "back to the future" and the way it was before we decided to have a PMO and began to manage projects with dedicated professionals. Or we will grow up as a profession and become an indispensable part of the staff functions of an organization just like Finance or HR. I'm not really sure yet where we are heading but we certainly better start to figure it out.

THE ROLE OF THE
PROJECT MANAGEMENT PROFESSIONAL

Even if PPM is eventually addressed more directly as part of the existing professional body of knowledge, a dilemma exists when project managers are called upon to address this gap in practice today. Given the state of understanding around PPM methodologies generally, they have limited firm methodological guidance to rely upon, making it hard to resolve this important executive dilemma.

The *Project Management Book of Knowledge* (PMBOK), which defines the professional body of knowledge for many North American practitioners and many other standard project management resources, suggest that the discipline of project management be applied beginning when a project is defined and approved. If we adopt the term "project management" in only this traditional sense, the dilemma of strategic project selection as discussed in the previous chapter would not exist; the project management practitioner would only be engaged once the project portfolio had been selected and approved for execution.

However, as a project management professional, this strikes me as a very narrow definition and risks irrelevancy for a professional partnership with executives. It has become clearer that the fundamental lack of appreciation within organizations for the value of project management could be attributed to what many see as a lack of strategic involvement and contribution. Nicholas Carr, in his now infamous HBR article "IT Doesn't Matter" (2003), makes a similar point about IT management—that ultimately IT will be seen as a utility and that investments designed to create competitive advantage out of IT infrastructures are doomed in the long run. His precepts and conclusions about IT could very well be applied to project management, if the content and context of our profession's potential contribution is too narrowly defined. If we only "manage projects," what's so strategic about that? Might this not ultimately render us more of a utility service than a strategic business partner to the enterprise? To avoid this fate, we must embrace what the online survey indicated executives want out of their investments in project management—project results that help execute their strategy.

For most organizations, this should be measured in terms of *business outcomes* rather than in the *methodology terms* that project managers are sometimes trained to think about. What an organization is looking for is not highly detailed descriptions of the steps required to achieve an outcome, but the outcome itself. While the methodology is important to the practitioner, it bores the business. A professional analogy may be made to accounting: when the CFO presents the financial statements to the CEO, there is rarely a grand and detailed description of how each entry in the GL was made and eventually accumulated to get to the end result. Nor is the theory of how specific entries are made of interest to non-accountants: it is the final output in which one puts faith and trust, and it is for this result that we hold the accounting professional responsible. The rest of us simply consume what they create.

Project managers must move in this same direction and begin to put more emphasis on outcomes, rather than on the interim steps, if we are to connect more with C-suite objectives.

What Organizations Really Want from Their PM Investments...

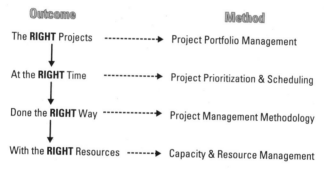

This is not a new observation and has been extensively discussed among project managers and executive sponsors in other forums. It also reflects my experience of trying to help executives manage their project management practices as end-to-end business processes that focus more on results than methods.

Yet few organizations actually seem able to address the four distinct outcomes shown in the above diagram successfully. Morris & Jamieson (2004), in one of the more recent books to emerge on enterprise project management, define the problem succinctly. At the outset of the book they write: *Projects and project management are often said to be important means of implementing strategy, but the way this happens in practice is rarely the subject of detailed review . . . there is little in the literature on how business strategy is translated into project terms. If we could understand better how business strategy can be translated into project strategy, project management's overall performance would be improved significantly . . .*

In the private sector, practitioners and academics have begun to address the challenge of project selection (a first and essential step in end-to-end project management) by adopting and then translating modern portfolio theory, as originally defined within the discipline of finance, into a project management context. As we have already seen, there can be no doubt as to the strategic value of getting this right, and most executives would see this as a high-value-adding contribution.

However, an important underpinning of the original theory is the normal priority placed on maximizing financial returns, while minimizing risk in the portfolio through balanced composition. This spreads the total risk of the portfolio over a broader number of investments that can generate the required rates of return. In applying the theory to project management, we can't ignore this point. It requires the development of decision support tools (normally in the form of some kind of consistent project scoring model) for portfolio selection.

Most of these scoring models emphasize selecting projects that offer higher financial returns as measured by traditional means such as return on investment (ROI), internal rate of return (IRR) or project payback. Other considerations might include amounts of available capital or resources, forcing the selection of a portfolio of only "affordable" or "doable" projects from the list of high-performing financial projects.

However, the emphasis remains primarily a financial one—not the same as being strategic. This is especially true in the public sector. In fact, others have previously questioned this whole notion of the profit-driven corporation and its potential negative impacts on society, government and the environment by promoting an increased need for corporate social responsibility (CSR) among global enterprises. But where would goals related to, for instance, protecting the environment, fit into a model of project selection that is essentially profit- or return-driven? Obviously, projects like this are not going to necessarily show a positive return and so would not be highly scored in most cases by scoring models with a financial bias.

So, if managing a portfolio of projects for financial gain is certainly not appropriate in the public and not-for-profit sectors and perhaps not even appropriate in the private sector, why do such models still exist and why are they so widely used? What should the right criteria be? Again, when we look to various sources of professional insight, there are not necessarily firm and clear answers to these questions.

As Morris (2005) points out, "Strategy management is a dynamic process; strategy is often not realized in a rigid, deliberate manner as planners often assume it may be." Similarly, projects that are good for the business need not arise only during an annual planning cycle and need not arise only within the ranks of executives. The process of connecting vision, strategy and proposed projects is dynamic rather than static and requires both a whole new level of thinking about the complexities of this issue and for project management professionals to be deeply involved and to completely understand the strategy-making process.

SELECTING THE MOST STRATEGIC PROJECTS

Even if an organization has a clear strategy that is clearly communicated (and, as we have seen, that is a big assumption!), this still does not address the practical complications of systematically choosing from among competing projects, all defined by their sponsors as "strategic" at the enterprise level. It also fails to address the primary issue of various projects being proposed on the basis of financial returns ("the business case") that ultimately will not be the most strategic set of projects from which to construct the final portfolio.

So we need to distinguish "managing multiple projects" and "program management" from "managing a portfolio of projects." In my professional experience, they are not the same thing, although they are often confused by practitioners and clients in practice. In a recent widely reported study, over 70% of project management practitioners surveyed in a variety of industries indicated they had implemented portfolio management in some form. However, more of the organizations in that survey incorrectly perceived portfolio management to be about managing collections of projects around a common theme (the more generally accepted definition of program management) rather than the correct interpretation of PPM, which defines it as "maintaining a balanced portfolio of projects through selection of the right projects and assignment of appropriate resources" (Morris, 2005).

As reported by a noted U.S. research firm (Gartner Group, 2001), "Ninety percent of U.S. companies do not employ a true portfolio management strategy." My own experience with clients echoes this reality, and the survey results shown below confirm it; clearly, most organizations are more comfortable with the simpler discipline of project management rather than the more complicated portfolio approach. The level of "don't know" (DK) or "none" at opposite ends of the graph shows the clear disparity in replies among respondents about their understanding of PPM in practice. In some cases, it is not even clear that organizations could answer the question about the degree to which they currently use PPM. Unless we attack this definitional issue as a profession, it will continue to impair debate on the relative merits of our responsibility and engagement around our role in PPM solutions.

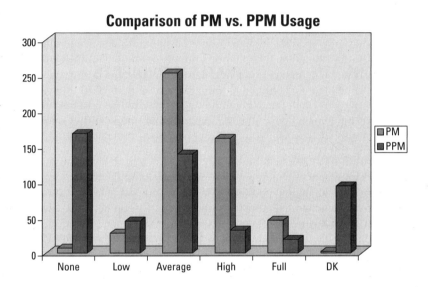

Comparison of PM vs. PPM Usage

For the purposes of this book, the term program management or enterprise project management shall be taken to mean managing multiple projects—as it involves modifying daily practices of managing a single project to the more complex but connected task of managing a group of concurrent projects with overlapping resource demands. This involves a higher degree of co-ordination across the enterprise in order to meet the "iron triangle" of cost-time and quality for every individual project. While this is orders of magnitude more complex than single project management, this is not the same capability as successfully managing multiple projects as a single portfolio.

Obviously, the more acute problem actually precedes the execution of the projects themselves. Before they are approved, projects must be formulated, defined and proposed to create the portfolio of projects that the enterprise believes will optimize its strategic outcomes. Then how do you choose the best projects to make up the final approved portfolio?.

Effective project selection is heavily dependent on creating and applying decision support tools to help executives in the enterprise make optimal project selection decisions based on a clear and cogent interpretation of the project's ability to impact the organization's strategy. This involves purposeful action by leaders in the organization to deliberately execute specific strategy trade-offs by making decisions about which projects get approved and activated, and in what order. *This is the most important and most strategic activity within any project portfolio management process.*

This approach means treating **all** projects, existing and new, as a single dynamic portfolio, drawing from a common resource pool, with the intent of maximizing business results. This way is emerging as the "better practice" and we will focus on it from now on.

UNDERSTANDING THE ORIGINS OF PORTFOLIO THEORY

Modern portfolio theory (MPT) was first defined by the Nobel Prize-winning economist Harry Markowitz in 1959, around the same time that Alfred Gantt was busy inventing modern project management while working in architecture, construction and engineering. Project management in business has always relied on the wholesale adoption of the founding principles in other fields being readily translatable into the modern organization. Now, this can be debated and may or may not be valid. For instance, if the origins of project management lie in architecture and construction, it is pretty clear when undertaking this task what it is that you are building. It follows an architect's plans and models, respects the building code and zoning bylaws, and must be built subject to engineering principles that are clearly established. The unknowns we have to deal with might involve factors such as the weather or labor strikes. But they are unlikely to involve much that is completely novel.

However, the same cannot be said of how we practice project management in the modern enterprise. The definition I hear associated with project management most often in a business setting is that it is work with definite start and end points that involves a "unique outcome," or similar language. In fact, we deploy project management on conceptual tasks precisely because they are not routine and well understood. We cannot borrow any theory and make it apply without significant changes: putting up a building is not like designing a new IT system.

Likewise, the notion of translating Markowitz's financial portfolio selection and capital allocation theory to business project management may suffer from similar limitations, even if the approach is no longer new and its theoretical base is proven. Having emerged mostly in the early 1980s among academics studying the project management discipline, this approach has a sufficient history in a professional context that can be examined in light of how the theory has translated into practice. But the primary value proposition espoused for the application of this technique, especially as expressed by Cooper (1997) and others around new product R&D, was to achieve the selection of an optimized portfolio of projects by selecting and managing all projects as a dynamic single portfolio. Confirmatory reports from practitioners suggest that while they consider the approach valuable, various aspects of the methodology become quite complex to manage and present challenges different from and often beyond the scope of traditional project management approaches.

As early as 1993, Cooper was able to report that there were many relatively divergent techniques that could be used to estimate, evaluate and choose project portfolios, but that many of these techniques were not widely used because they were too complex and required too much data manipulation. Or maybe they were just be too difficult to understand by practitioners and so may not have been adopted into use.

But none of these basic theoretical approaches can ever be successful. Why? Because ultimately, strategy is not completely financial in nature; nor is risk, and so the assumptions inherent in this approach negate the complexity of most organizations' true context. And since we can *only* choose from among the projects presented to us for scoring, rather than objectively comparing the portfolio of projects currently presented against the total strategic gains required by the organization to fulfil its plan, we may be needlessly assuming that strategic success will occur if we execute the topmost projects produced by this model. But what have we potentially missed by using this approach?

Without the ability to specifically measure the strategic contribution of any single project against another, and without the ability to embed strategic and non-financial targets that matter for the organization as a whole into PPM scoring models, it is not possible to select a portfolio of projects that optimizes strategic outcomes and intent. While private-sector business strategy often has a profit-making focus that makes the problems identified previously less acute for them, there is still value in refining PPM practices for this sector, to incorporate non-financial strategic outcomes into project scoring tools.

Of course, any approach to portfolio management involves scoring projects. This requires an improved tracking and control system that takes into account the needs of executive decision-makers and real-time performance management of the active portfolio of projects against proposed new projects. It requires the use of post-project audits and reviews, including managing organizational knowledge about practices that did and did not contribute to achieving individual project outcomes and the organizational strategy. Comparisons of this data can then be fed back into future revisions of an organization's chosen scoring model to refine its predictive ability. Longitudinally over time, this should improve actual project outcomes for the organization and is another potentially huge value-add of PPM. However, this requires the consistent application of post-project reviews within an organization.

Yet, research shows post-project reviews are often skipped, creating another gap in practice to the implementation of a fully functioning project portfolio management approach. Not doing these reviews means you lose the value of "lessons learned," and can keep your organization in the fog.

ON-STRATEGY PROJECT MANAGEMENT: THE FOURTH DIMENSION

After careful reflection about the extensive discussions with both executives and senior practitioners that occurred during this research, it seemed to me that our profession could benefit from the addition of an "on-strategy perspective" around traditional current project management practices. What does this mean in practice? It means putting strategy at the center of everything we do.

For instance, it could mean putting into action specific methods (in our examples, for the selection of projects) that generate the most strategic outcomes for your organization. This implies that "better practices" will vary according to each organization's context and objectives: there is no single set of rigid "best practices" appropriate for every organization. Project management practitioners need to be willing to move beyond the traditional iron triangle and make sure the project work they undertake truly is strategic.

While I do not wish to lose sight of important considerations such as on-time, on-budget, and on-scope, infusing and connecting considerations about how to accomplish these outcomes strategically to benefit the execution of an organization's intended strategy is the right thing to do. This creates alignment, that defining outcome we talked about at the outset of this chapter.

During a pilot study conducted as part of a joint effort with another consulting firm, we surveyed a fairly representative senior group of

Adding a Fourth Dimension to the Iron Triangle:

On-Time

On-Strategy

On-Budget   On-Scope

**Competent People + Good Process + Clear Strategy
= Extraordinary Results**

informed project management professionals from across Canada. While I made no attempt to validate this statistically, it is likely that this group would also be representative of any senior group of project management professionals in North America (or around the world).

These respondents reported significant knowledge about project management (PM), but fewer reported accompanying or even similar in-depth knowledge of project portfolio management (PPM). The contrast here is notable and may be partially illuminated by looking at some of the specific results:

Of those respondents with PPM knowledge (47.5%), virtually all report it as having a high complexity.

Half of all respondents report having no knowledge of PPM, despite having high awareness and experience with PM, and 47 (59%) have no PPM process in place within their organizations today.

Of the respondents surveyed, 96% reported they "need to learn more" about PPM.

Of those respondents who are using PPM, there was only limited endorsement of the process as being effective, and most reported a neutral to slightly negative perception overall of the results of the methodology in practice.

These trends in knowledge and practice are easily compared using comparative tables showing the relative differences in reported knowledge levels between project management and portfolio project management methodologies (see below).

The shape of these two graphs (which have the same scale) tells the story descriptively and completely: respondents self-report considerable knowledge about project management practices but not a correspondingly high knowledge of PPM practices. Now, it is fair to say that these early results may have something to do with the relatively recent nature of these techniques making their way into the professional dialogue and their absence in many current sources of professional practice. However, this survey does verify that, as previously suspected, knowledge of PPM practices is low in the profession generally. As an executive, this means that you cannot assume that your most senior project management professionals necessarily have a command and grasp of PPM or that they are necessarily skilled in its implementation. Or that they understand the connection between on-strategy project management and portfolio management practices.

Similarly, among professionals who have bravely attempted to implement PPM in practice, often with few sources of professional guidance to follow, there was a mixed degree of satisfaction with the results of their efforts. Most reported neutral to negative outcomes. This confirms my initial view that there were probably issues present when trying to implement the current methodology even though some professionals reported good results. So, if you are an executive who has had a bad experience with PPM, I ask that you refrain from pre-judging the process and instead focus on what went wrong in the implementation you were involved with and how those issues could be addressed. Otherwise, there is always the risk that early adopters trying their best will actually be blamed for these failures and risk the organization never attempting to move forward in this area ever again. That is a most unfair and unfortunate outcome should it occur.

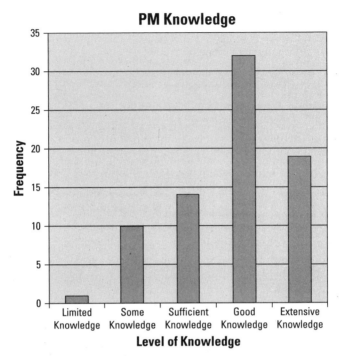

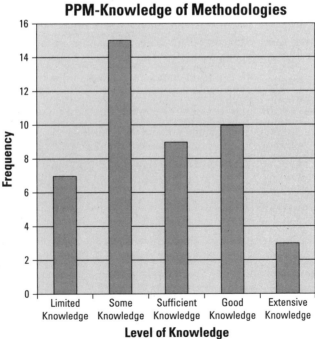

Pilot Study: Graphs Comparing PM and PPM Knowledge Levels

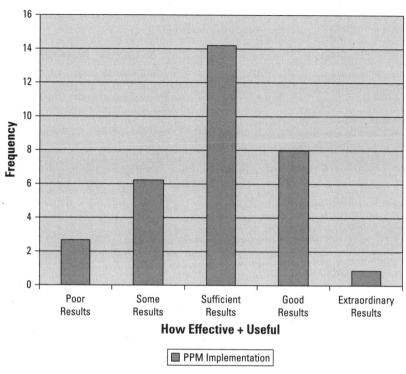

Pilot Study Table—Results of PPM Implementations

THE FINAL WORD IN THE
REAL WORLD OF PRACTITIONERS

I attended a conference in Toronto some years ago where I led a plenary session to share the results of this survey followed by roundtable discussions. Keep in mind that this is a well-educated, experienced group of peers; here are a few of their comments:

"I've never understood PPM or Enterprise Project Management—what is all this stuff going to do for me, anyway?"

"PPM wouldn't work at my company unless our executives want to give up the power to decide which projects to do ourselves . . ."

"When we started off with PPM, it was too complicated and the PMO ended up being seen as less value-adding than before when we were only internal project management experts."

"Our company doesn't seem interested in strategy—only profits."

"We've used PPM for two years and it seems to be working but it has made us focus mostly on fewer projects."

"The problem with PPM is the same problem as with our profession in general—we only half think things through before we start talking about them and then doing them—stupid!"

"Our PMO started trying to implement PPM but it seemed to mostly add work and not many benefits except we have started to pay more attention to risk-managing our big projects."

Generally, the commentary (including conference participants who had not filled in the survey) suggested the following conclusions about PPM:

1. It was not extensively in use among participants generally.
2. It was perceived as overly complicated and not as developed or articulated as it should be in order to be practical.
3. In its present form, it seemed to offer limited benefits.

As a result, many of those in the room were resisting PPM because it could risk their social capital or reputation within the company. This may be as much a result of the way PPM was being implemented as it was PPM techniques themselves. However, it did make clear that if the project management profession was to become more strategic, and implement PPM effectively in partnership with executives, changes were going to have to be discovered that would enable better outcomes. During the pilot study, I also discovered some critical information about strategy and, since PPM is by definition a "strategic system," its relevance helped open more doors to the possible solution to the problems I had already started to discover.

## CHAPTER THREE

Navigating Through Foggy Strategy

Most organizations have a stated strategy. It may be clear or not and it may, in hindsight, turn out to be the right or wrong strategy. It may not be published, but simply shared among colleagues informally. But no matter how you cut it, most organizations have one.

And organizations typically have a management methodology for project work. While the methodology may be simple or complex, designed to handle single or multiple projects, its primary focus is still on managing the work of projects themselves. So what is missing is the process to help organizations translate strategy into the right project selections and resource assignments, even before the process of managing projects begins.

I have found while consulting with various kinds of organizations around the world that most actually have *strategic intentions* rather than *defined strategies*. Motherhood statements such as "become a customer service leader" or "be a low-cost provider" are examples of strategic intentions. These are intentional more than specific, because they are not measurable: you never know when you have reached your goal! For instance, asking how low costs actually need to go in order to be globally competitive would be the right question to ask. You can then institute specific cost targets and decide how best to reach them. Now we have moved from intention to action and that is what good strategy is all about—it enables concentrated organizational action.

If we return to the generic PPM process diagram found in Chapter 1, the importance of clear measures relates to the circled steps shown below. Without the structure of strategic measures, developing an effective scoring model to select projects and set priorities is impossible.

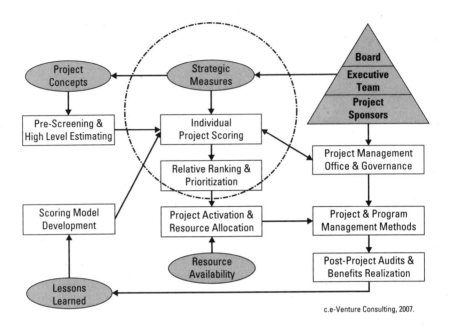

c.e-Venture Consulting, 2007.

Vague strategy evokes a common response in organizations I have worked with: people wait for strategy to be made clear. They want the *intentional* to become *practical*. This rarely happens, in my experience; worse still, as we wait, time becomes our biggest enemy and the motivation to move forward is lost. This results in lost business opportunities and employees become less productive.

We need not wait for strategy to become clear, but rather we should *discern* how we might act to make strategies clearer and contextualize them with what we *do* know. To accomplish this, we need a clear strategic context at the highest (usually global) level; it can then be adopted into guidelines for local action. I call these *decision filters*. When strategy is internalized and understood, individual employees can use this information to make decisions. By improving results one employee at a time, collective action is improved, resulting in better overall results. Anyone who has worked at GE can tell you about this: their clear understanding of the link between strategy and results is legendary. They articulate all corporate strategies with critical measures (e.g., be first or second in your industry, or you become a target for divestiture) and then make these highly visible to employees at all levels. This creates alignment between employee efforts and expected results.

Can you imagine if we organized the running of the Boston Marathon the way many organizations set strategy? You set off on a race with the specific destination unknown (just "run the course"), and it's not clear how long the race is or where the milestones occur. So you simply set off and run—along with a bunch of other competitors. Who wins? When? How? What are the rules about staying on a particular course versus setting your own course? If you win the race another way (say by hijacking a golf cart), is that okay? Or does how you run the race matter as much as simply finishing the race?

Sometimes corporate strategy seems to be organized in a similar way: it's well-intentioned but lacking in specifics that would enable us to clearly determine when we have won the race! This makes it hard for individual employees to feel empowered to act, despite their best intentions to try and make a strategic difference.

This was confirmed in the early pilot study we conducted. The study helped me explore the nature of existing PPM processes, especially their current prevalence in practice. But the lack of clarity around organizational strategy is contributing to a breakdown of PPM methodology in practice. It is hard to be deliberate in the selection of a strategic portfolio of projects if the strategy itself is unclear. And the strategy must be measured specifically and deliberately to create organizational alignment, as shown in the diagram below.

The Effects of Organizational Alignment

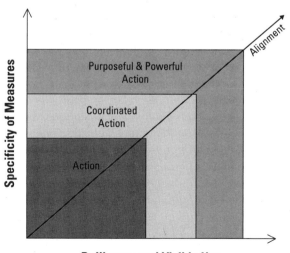

As the specificity of the measures used in your organization to measure the strategy improves, so does an employee's ability to link their individual actions to strategic outcomes. Similarly, the more visible this action—i.e., employee communication strategies, linking measures to compensation or performance review systems, quoting measures in job descriptions—the more real the notion of linking strategy to action becomes. These two changes, made simultaneously, improve alignment of individual effort to organization strategy—a powerful goal of most executive suites.

The power of this concept was confirmed by comparing the "clarity" of the organization's strategy with the notion of "well understood" (see the graph below) and establishing no correlation between the two. Furthermore, I discovered that few organizations seemed to have a set of strategic measures designed to help them determine how well they were executing their strategy. Since, as we now know, projects are the building blocks of strategy execution, without these critical measures, how are organizations actually ensuring that the project portfolio they have is strategic?

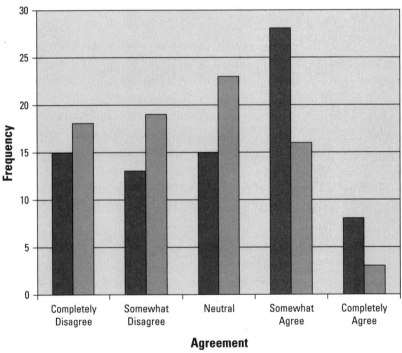

Research Results on Clarity of Strategy

Unless an organization's strategy is clearly measurable, it is unlikely that it will be understood well enough within the organization to be used as a guiding force for strategic project selection. This is a critical breakthrough in terms of what is currently missing in PPM methodology. Since measurement is often the means by which we provide clarity about fuzzy or abstract concepts such as strategic objectives, this failure by organizations to make their strategy measurable is important.

Data suggests that respondents may feel that their organization's strategy is "clear and concise," but this does not necessarily translate into "well understood" throughout the organization. So what are the implications of this when we try to connect projects to strategy? The pilot survey suggests that a PPM methodology is only used to relatively rank and execute projects on some basis other than strategic priority. Because the strategy is not clear, we cannot say with certainty that any one project is more strategic than another. If these responses are generalized, it would suggest that those using PPM today must be doing so with only a partial understanding of their corporate strategy; thus, they may be getting only suboptimal results from the process anyway. As shown in the diagram below, that is the central mission of the research undertaken for this book—to create optimal results from a PPM process by clarifying the link between strategy and project selection.

For project management to become a truly strategic contributor to organizations (as measured by improved strategy execution), we must be ready to engage strategically!

Voices of Experience:
CEO of a Major Outsourcing Company

Laurie is the CEO of a major business services company that provides telecom and IT outsourcing to clients throughout Canada and the eastern United States. Much of the firm's work is project-based. We got her to stop for a few minutes and talk to us about how she sets strategy and what she expects in terms of its execution.

Why is project management important to your business?

As an outsourcing company, project management is the business. We make very specific, concrete promises to our clients in our sales proposals. These proposals eventually become contracts that will guide our people and clients towards implementing a new business relationship. This makes project success a high-stakes game for us.

Does this change how you do projects?

Compared to what—how other companies do it? What always interests me about project managers is their belief that some projects simply fail because they are bad projects. There is no such thing as a bad project— only bad managers—either the business leader created something that could not be done or the project manager assigned to it failed to execute. We cannot fail to deliver service to our clients because that is what they pay us for and what they are expecting from us. This means being very careful to structure our projects in great detail and obviously lots of measures. This keeps both us and the client honest about how we are both doing. Without these service level agreements we would be lost.

As a CEO, what do you expect from your project managers?

Results. No matter how good a project looks on paper when it is planned, it's ultimately got to work. If it doesn't, the effort doesn't matter. This means that project managers and business leaders must both be willing to be accountable for measurable results and this is how we structure ourselves.

BALANCED PERFORMANCE MEASUREMENT AND MANAGEMENT

After more than ten years' experience using the Balanced Scorecard (BSC), I can attest to its value as a tool to measure and map organization strategy in concrete terms. Of course, if this outcome is achieved, it follows that the projects selected should also be more strategic. I am not proposing the use of the Balanced Scorecard specifically, but that *all organizations use some form of clear and complete strategic toolkit that enables them to express a set of balanced performance measures linked to their planned strategy.*

For those not familiar with this tool, the BSC focuses its efforts on helping organizations develop a strategy across four standard dimensions or perspectives: financial, customer, process and people. The notion of balance is important within this framework—a complete strategy must address all four. Traditionally, financial measures were the measure of performance in most organizations. They are the easiest to identify and deal with, but are lagging indicators of what has already happened, not of what may come. In fact, this mirrors what we see in traditional PPM methodology: the dominance of financial measures as the ultimate success criteria.

The BSC requires leaders to think more conceptually and broadly about their definition of organizational strategy and performance measurement. Thus, leaders are challenged to ask themselves basic questions regarding their intended strategy in each of the four areas, as represented in the figure below.

The Balanced Scorecard Perspective

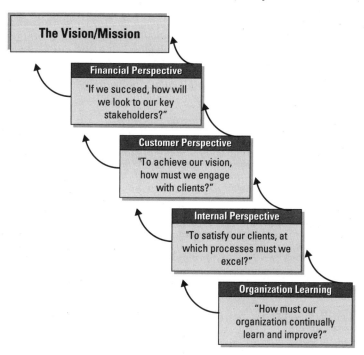

This method is generally successful at determining what future activities will likely have the most impact on the organization's strategic objectives (strategy mapping) and making sure that initiatives and projects exist in sufficient quantity in all four dimensions to accomplish the intended strategy. Since the strategy is mapped with the intent of unlocking the potential of the tangible and intangible assets of the firm (Kaplan & Norton, 1996), often this results in a focus on the most highly strategic tasks that will provide maximum cumulative value for the firm. The strategy map makes the steps along the way quite clear and helps leaders identify dependencies and important interactions among the tasks at a high level. Specific measures and timelines create a sense of urgency. This permits the organization to closely track its strategic accomplishments, acting as a bridge between strategy formulation and execution. It also means we can connect project outcomes to *specific, measurable strategic outcomes.*

One of my clients, who is with a major global bank, did this by creating a "return on strategic measures" (RSM) approach. She defined this as the cumulative total benefits of all projects, compared against the total targeted improvement level required to meet the targets set for the company's strategic measures. This enables her (the CIO) to quote the specific percentage of the contribution of each project to total required corporate results. Tracking this has been quite insightful.

This powerful application of PPM has significantly enhanced conversations with her executive peers about how to prioritize and manage projects with competing needs, costs, timelines and resources. This is the kind of powerful breakthrough that is possible when we can more directly connect project outcomes to measurable strategy.

BUILDING A STRATEGIC PROJECT SCORING MODEL

So what must change in order for organizations to consistently prioritize projects strategically? To start, we need a systematic scoring model that enables comparisons about the relative value of different projects to be agreed on among executives, sponsors and project managers alike. While theoretically easy to accomplish, my experience suggests it is often hard to put the theory into practice.

Measuring Projects for Their Strategic Effectiveness

First off, it must be possible to assess the individual strategic contribution of any one project in relation to the organization's intended strategy. This implies a means of measuring the strategic contribution of each project at the individual project level. If your organization is using the BSC, then you have a solid foundation on which to base these decisions at the enterprise level. Connect project outcomes to their measurable impact on the strategic measures of the organization, across the three broad dimensions of strategy (for example, people, process, technology). Or use the four dimensions of the BSC, if that is your chosen tool. Or use a set of key performance indicators (KPIs) established by your corporate parent, board or executive team. Regardless of the system you use, it must render the expected results in measurable terms, or you cannot connect projects to strategy.

For users of the BSC, you would have specific strategies mapped. Associated measures and targets help you know when you have "crossed the finish line" in each of the four dimensions. This provides specific, measurable objectives and can help project managers align their projects to those critical results. By embedding enterprise-level strategic objectives into your project initiation/project proposal system, you create a scenario that demands that those who conceive projects must be aware of the value of the project in strategic terms from the outset.

This method also helps in setting priorities; each project naturally has a higher or lower level of contribution to each individual dimension of strategy, depending on its make-up. Using the BSC, a project could contribute in one dimension or in more than one—or even in all four.

Find Strategic Balance Through Project Allocation

Another critical aspect of the methodology is balance: an organization can create a sufficient portfolio of projects to ensure that its stated, measurable objectives (linked to its strategy) can likely be accomplished with the mix of projects that are being proposed. This eliminates the risk of only picking projects that execute on, for example, financial strategy but not on the other dimensions. The BSC helps identify any gaps before determining the final mix of projects in the portfolio—a powerful and more complete approach to this all-important task.

To put this into practice, it becomes important to find a method that is relatively simple for making these comparisons among projects based on their measurable strategic contribution—for those proposing projects as well as those selecting the approved projects.

Traits of an Effective Scoring Model

The tool to accomplish this is a scoring model—an essential ingredient of any approach to PPM. Again, the purpose of this book is not to innovate around what we already know about PPM, but to provide useful and accurate guidance on how to create and operate an *effective* scoring model. In our practice, we define the term "effective" as a tool and an accompanying process that both deliver on the required result (selecting the most strategic projects for the portfolio in rank order), while balancing the outcome with the need to limit complexity and costs of implementation of the tool and process itself. This is derived from the discussion about effective versus efficient process designs from Chapter 2.

In the following example, the organization has a series of strategic measures and indexes that it has established as critical to its strategy. They are shown across the top.

You then assign an arbitrary rating scale with four intermediary points (2, 5, 12, and 25) so that each point approximately doubles the weight of the previous one. This is known as a non-linear scale, and it is a deliberate and important choice. The intent is to create discrimination between projects and perhaps to graph these to make them visually obvious; therefore, regardless of the specific scale you use, it must clearly separate high-scoring from low-scoring projects. The numeric scale itself is unimportant and could be any set of numbers you like. The purpose remains the same: a scale that everyone agrees will define the contribution of projects differently, forcing the decision-makers to discriminate among them and help with project selection decisions.

Each project (in this case A through H) is then scored individually, based on the connection between the specific project outcomes and the four dimensions of the project: impact on People, Customers, Service, and Finances.

LEGEND				(Balanced Strategy Measures or KPI's Unique to Your Organization)				
(Degree of contribution to strategic measures of your organization)		Project	FINANCIAL MEASURES	CUSTOMER MEASURES	PROCESS MEASURES	PEOPLE MEASURES	Total
Very Direct	25	Project A	2	25			27
Direct	12	Project B	25	12	2	5	44
Indirect	5	Project C	0	2	25	12	39
Weak	2	Project D	0	0	0	25	25
None	0	Project E	5	12	0	0	17
		Project F	5	0	0	0	5
		Project G	0	0	25	0	25
		Project H	2	0	0	12	14

Scoring Model Output

To avoid creating a false sense of security around the mathematics of the model, the points on the chosen scale were then assigned labels (*weak*, *indirect*, *direct*, and *very direct*). These are designed to be an expression of how much measurable contribution an individual project makes.

Obviously, this still involves executive judgment in terms of the scores themselves; but establishing a consistent approach to project scoring is a good first step.

When working with clients to develop these models, I always point out the necessity of transparency in this work. To garner organizational buy-in, especially from senior executives, the scoring model must be simple and clear. Each measure used must have associated definitions that help those creating projects to understand how they will be assessed. An unclear scoring model will seem arbitrary and will only generate suspicion and anger.

Promote Clear Communication Through Thematic Analysis

Another key part of the success of this type of project scoring tool is generating consistency of interpretation among executives when project proposals are created. Beyond making sure the model itself is transparently clear, I offer a few techniques to our client executive teams on how to learn to rate proposals in order to create a consistent scoring model.

Often within organizations, language gets in our way: the CFO may speak of the "value of a project" in very different terms from a social services case worker in government. Their commitment to their own disciplines, to their professional experiences and mandates, and to technical interpretations of concepts, may make it seem impossible for them to define a project. Yet, in my experience, they can. While challenging to institute, it involves borrowing a simple research tool from the social sciences called *thematic analysis.*

Take the CFO and the child services worker: as they describe *how and why* they would rank a project a certain way, there are bound to be commonalities. We call these *themes.* For instance, we might be able to establish that each of them would agree that a project that is closer to the client (i.e., the end user of the accounting system, or the child at risk) might be considered more "direct" in its contribution than a project that is remote from the client. We might further get them to agree that projects that are of higher value are able to more clearly discriminate their contribution to supporting work flows that are more complex versus less complex in nature. To accomplish this requires that for a period of time you *listen* to the descriptions of projects and the rankings and, using thematic analysis, discover the commonalities in the language, descriptions and approaches used among senior executives. When this is done a sufficient number of times, it is possible to begin to reflect back *generic* but substantive descriptions to attach to words like "direct," "indirect," and "weak" that all executives, regardless of their function or discipline, can likely agree to and apply.

Applying the Model Retrospectively

Another tactic that I have found helpful when implementing PPM for the first time in an organization is to encourage executive teams to first define the tool and then to experiment with its application to the prior year's project proposals and selection decisions *retrospectively as a learning exercise.* This accomplishes two objectives: it is a low-risk environment; the decisions have already been made, usually the previous year. This eliminates a certain amount of jockeying and politics. Second, executives are able to develop a consistent and internalized understanding of the definitions of, for instance, *weak* versus *indirect*, by looking at projects that are generally well understood because they have already been reviewed and approved. This tests both the validity of the scoring model and their understanding of it in practice.

This technique helps when the scoring model is then applied to new projects, where the vested interests and risks associated with the decisions are higher. The team has already begun to trust the model.

Since the scale or values of the scoring model are not as important as its consistent application to ensure that selected projects have a more tangible, direct contribution to the organization's strategy, there is a high degree of latitude in the actual design of the scoring model itself. What is important is that executives see projects as having a strong alignment to the organization's measurable strategy. This can be depicted by connecting the size of a project's bubble (based on the labels *weak, indirect, direct* and *very direct*, each corresponding to a particular circle size) to the organization's strategy.

This modification to more finance-oriented scoring models incorporates specific strategic measures, which in any organization may be based on the BSC or other measurable strategic imperatives of the firm, but still includes traditional financial measures of project performance. These are simply incorporated into the financial dimension. However, it is the essential element of balance across the many dimensions of strategy rather than relying only on project-based financial measures that is the breakthrough.

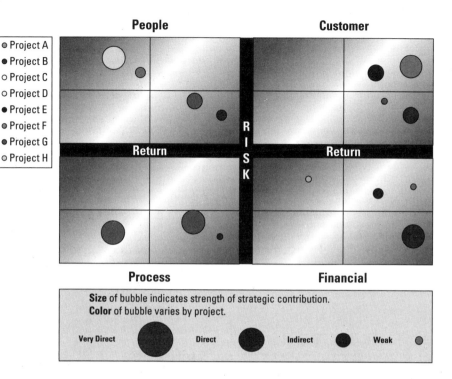

| ◉ Project A |
| ◉ Project B |
| ○ Project C |
| ○ Project D |
| ◉ Project E |
| ◉ Project F |
| ◉ Project G |
| ○ Project H |

People   **Customer**

Return   R I S K   **Return**

Process   **Financial**

Size of bubble indicates strength of strategic contribution.
Color of bubble varies by project.

Very Direct   Direct   Indirect   Weak

Within the public sector, this change in methodology becomes even more critical, because project scoring can now accommodate the more complex and non–profit motivated strategic objectives often found in this sector. Using the BSC (or any similar balanced set of performance objectives), leaders can reflect the complexity of strategy found in legislation, government social policies and practices, and other factors that often cannot correlate to the simpler profit-based strategies found in the private sector. Since this is the context in which a public sector project manager must establish project priorities, it is essential that any PPM methodology used in this sector be able to accommodate this additional level of complexity.

However, there is a *critical conclusion* about the assessment of strategic return for projects in an organization: there is no way to standardize this to a fixed formula that can be universally applied. The definition of each organization's strategy, strategic measures, and associated definitions of how it assesses individual projects will be unique. In fact, the approach is most effective when it is customized within the specific context of your organization, since executives will devote time to constructing and using a tool for which they feel a high degree of ownership. So while it may be more work, it is vital work and cannot be done by buying an off-the-shelf PPM approach and assuming that any vendor's definition of factors to define project contribution will be accurate in your own organization's unique context.

There is another important shift in understanding created by combining the two methodologies: we are now able to assess an individual project's potential and its overall strategic contribution rather than comparing them relative to each other. The difference is critical: the way to approach project scoring is relative to a *fixed target* that our strategy says we have to meet. For instance, if we talk about impact on internal process efficiency, we can measure how much any specific project will contribute to our planned level of performance. This moves us away from using artificial or ill-defined measures of "project success" to a strategic interpretation of the value of a specific project in a specific portfolio at a particular point in time.

BENEFITS AND COSTS OF
REVISING YOUR CURRENT METHODOLOGY

The findings from various case studies we have conducted, three of which appear later in this book, support both the usefulness of this approach to project scoring and the benefits of strategic clarity that methods like the Balanced Scorecard bring to an organization's understanding of its strategy. It has now become clear that the application of this methodology is seen by executives as strategically optimizing their organization's portfolio of projects over the long term. And project management practitioners like the approach; they find it logical and practical, as it helps reduce the complexity and anxiety previously attached to implementing PPM.

This has been confirmed not only by workshop participant feedback, interviews with key executives and the limited use of specific questionnaires comparing pre/post intervention attitudes, but also in the longitudinal results of organization performance versus project completions. In other words, as projects are completed, you can ask: did they deliver their intended result? Comparing the planned project results with your achievement of strategic outcomes expressed in the measures makes the methodology useful as a corporate performance management tool.

Non-financial impacts of projects are often described as "vague" or "unmeasurable." Significantly, with our approach to project scoring and selection, this is not the case. Our case studies provide evidence that after being exposed to the methodology, participants gained a superior ability to connect project outcomes to strategy and reported a clear sense of which projects were "strategic."

It is also clear that study participants are more comfortable with taking previously ambiguous strategy and turning it into measurable outcomes that can be linked to project outcomes after they understand how projects are going to be scored using the PPM framework. This is powerful and goes well beyond what most organizations are doing today. This works especially well in the public sector.

While no assurance can be offered that my proposed framework and tools are the perfect solution to resolve this challenge, it is clearly superior to the status quo financial models currently in use and provides a starting point for the use of balanced scoring models as an excellent addition to existing strategic project management tools.

A final question remains: is the work to implement this revised PPM method worth it for the benefits achieved? This issue is more complex than the previous two. While my research clearly shows benefits from the new methodology and validates its strategic contribution, there are costs associated with this methodology that are not inconsequential. I surmise from the sample of organizations that I have worked with that the costs and complexities of the revised methodology will initially be limited to larger, global organizations, and very large public sector organizations where the sheer volume and complexity of the project prioritization process demands an effective method. For organizations of this scale the benefits outweigh the costs of this approach. However, as interest and proficiency grow within the profession, future research is bound to find ways to make this method as cost-effective for smaller organizations as it would be for larger ones.

During the research, study participants mentioned other potential benefits. These cover a broad spectrum of practice areas in project management and are noteworthy because in the long term, if they continue to consistently emerge in our future case studies, they will become additional value-adding outcomes that arise from our PPM framework. These are summarized in the table on pages 55-57. Each includes a selected comment from a case study participant that has helped me better understand what does and does not work practically in organizations trying to implement PPM.

Based on our work with this methodology over several years and with hundreds of clients, our clients consistently report these as benefits. While it's clear that not every executive sponsor, project manager or staff member may see these in exactly the same way, there is clear value attached to the effort required to implement PPM in your organization, and I believe that more and more of these benefits will be confirmed over time.

PUTTING THE PROPOSED CHANGES INTO PRACTICE

Adopting this change in methodology removes PPM from the purely financial realm and ensures that organizations do not pick only economically efficient projects. For instance, for projects related to longer-term investments in process or people management effectiveness, there might actually be lower than normal financial scores when these projects are compared to other proposed projects. Does this suggest that an organization should not do them? No. But if your organization's current scoring model puts too much emphasis on internal financial measures derived from the business case, then this would be the likely outcome, though obviously not the best strategic result for your organization.

Adding new dimensions and the notion of balance will improve project selection throughout your organization—and, ultimately, will give you a different final portfolio than if you had focused more narrowly.

Of course, to implement a PPM scoring model like this requires that the nature of how the organization conceives documents and proposes projects for approval will change as well. This means integrating how the business case for any particular project is built, so that it no longer is based solely on financial returns.

When the business case for the average project contains an elegant strategic analysis, we feel renewed confidence in a system which, like it or not, has become in many organizations akin to writing fiction. Given that many non-financial (but also non-strategic) factors rank in the project selection processes of most companies (who is proposing it, how much it costs, etc.), it is imperative that when we tackle the implementation of PPM we also address the organization's business case template and bring it into alignment with our new view of PPM and enterprise project selection. We will discuss project submission further in Chapter 5 when we talk about the role of the Project Management Office (PMO).

Finally, combining these two approaches (BSC and PPM) offers an organization the opportunity to establish a common language and a common set of measures to determine the scoring of projects. One of the hallmarks of the Balanced Scorecard methodology is its insistence on measuring strategic success at the organizational level and then

Table 3-1 – Ancillary Benefits Summary

Interview Comment	PM Practice Area	Possible Impact	Possible Benefits
"I now begin to conceptualize projects from a strategic perspective from the outset rather than waiting till the project has been planned and then considering its impact"	Project Initiation	It would appear that participants begin to initiate projects differently once exposed to PPM. From the outset they begin to reflect on the strategic contribution of project ideas before investing time & money into defining the initial business case as might have previously been the case.	Using the strategic filter at the project initiation stage may reduce the investment of resources to define projects that ultimately are not sufficiently strategic to proceed.
"The flow of projects has been reduced to a manageable level now…thanks!"	Project Approval	By forcing capacity planning into the project activation process, organizations are forced to consider not only project priority but their ability and the availability of resources to execute projects more carefully.	Probably reduces spurious projects but also ensures that projects are activated only when resources are available to do so regardless of approval or priority.
"Communicating business objectives to my project teams is much easier with this new method although it does take more time initially to put in place"	Project Communication & Reporting	Commonly, participants involved in project leadership often detect an improvement in their ability to communicate strategic business objectives to their project teams. Also those on projects report that because the strategy statements and measures remain constant but the projects they work on may change, there is limited re-learning required when moving between projects.	Since business objectives get clearly and consistently communicated, this may allow staff to more quickly move between projects and realize the connection between their activity and strategy improving productivity and clarity of mission.

Interview Comment	PM Practice Area	Possible Impact	Possible Benefits
"As the CIO, I no longer argue about getting approval for more resources—I now argue about how fast I can actually add them!"	Project Staffing	Traditionally, staff organizations have been viewed as a cost center and investments in these functions were controlled or limited by this perception. In addition, often the leaders of these organizations found themselves perpetually under-staffed and always feeling like they had to defend current performance (often seen as inadequate but mostly because more work was given than resources were available to complete it) while seeming to always ask for more staff.	Since PPM drives to a strategic view of project approvals rather than a resource-constrained view, very different approaches to these kinds of decisions result. Most often, resource loads can now be directly predicated on the approved project list ensuring more balance between supply and demand in these functions and improving morale and performance.
"Since I cannot be everywhere, every time a decision has to be made, I now rely on the proxy measures to guide my staff to make the right call...."	Project Quality	Projects consist of many hundreds of thousands of individual decisions, ideally aligned to create the ultimate output. However, issues can arise when project staff do not share a common set of parameters to make decisions. Historically, this has sometimes led to project leaders trying to centralize project decision-making but at the detriment of speed of execution. PPM now offers an alternate approach to generating consistent decision-making frameworks that are clear and measurable.	When decision-making becomes consistent, exception reporting goes down and productivity goes up. The ability of project team to make more right decisions more often is a known factor that drives up project quality.

"The relationship strains of the budget scrum are no more!"	Budgeting	In many organizations the planning and budgeting cycles collide in unproductive ways. Often the two are actually in contention as budgets drive strategy rather than the other way around. This was frequently identified both in the public and private sector case studies by executives who did not always feel previous methods used were all that productive.	Having the ability to reduce the interpersonal strain of executive debate by depersonalizing it likely renders the process more "user-friendly" and, although the subject matter remains the same, the change in approach may translate to executives feeling better about the PPM method as a tool to have these trade-off discussions.

benchmarking this to internal and external performance goals. This same approach benefits PPM by reducing the amount of perceived scorer bias when projects are presented. This will likely have a significant impact on leader behavior within an organization and affect decision-making practices. It is particularly clear that the traditional discussion during the annual "budget scrum" would likely move away from the question of who proposed a project, or who controls the budget to manage these projects, and so on. Instead, planning meetings can focus on more essential questions.

On the strategic side, the Balanced Scorecard or a similar strategic framework that is measurable forces executives to clearly state their strategy and to develop specific measures associated with these goals. If an organization is going through this process for the first time, as we saw before, overall organizational clarity of the mission/vision and strategy of the organization increases at every level.

Meanwhile, the project management profession has noted and complained of inconsistencies in corporate strategy and the resulting inability to address this challenge by making direct strategic connections between business vision and project strategy. Therefore, project management professionals will likely be quick to see the benefits inherent in the integration of these two methods—and "lead the charge" to implement this new model.

## CHAPTER FOUR

Managing Project Risks, Returns and Resources to Maximize Benefits

The entire project management profession is currently talking about benefits realization: the notion that when projects are complete, they actually deliver on their promise and provide a return. Executives, in turn, talk about getting "bang for the buck," making sure the funds invested in project management actually increase shareholder returns. A lot of what has been learned over the years can help both parties achieve this outcome.

We can also tie existing enterprise risk management (ERM) strategies into our PPM process to ensure that, while getting high returns, we balance this with the identification and mitigation of both internal and external risks.

MAXIMIZING YOUR PROJECT MANAGEMENT EFFORTS

PPM is a relatively recent methodology that at first may seem like a simple extension of multiple project management methods. But the complexity of truly strategic PPM practices is much harder to implement than simple enterprise project management practices, and the two should not be confused.

Over the past five years, I have had the chance to ask hundreds of practitioners and academics in various professional settings about their views on PPM. Their individual comments support the conclusion that there is often limited understanding of or interest in the implementation of PPM because of the perception that it is too complex and offers little gain for most organizations. Clearly, I do not share this view or I would not have written this book. Properly implemented, PPM provides the capstone of a proper hierarchy of project management maturity. In fact,

Project Management Hierarchy of Complexity and Benefits

as an organization you can consider the maturity of your project management practices as a hierarchy (as shown below). It moves up in increasing levels of complexity but with increasing benefits. Strategy execution is what most executives really want the focus to be on. So this chart helps us figure out where we are currently and what kinds of project management practices we need to focus on to move to the next level of sophistication.

Most project management maturity models such as, OPM3 (a standardized maturity model supported by PMI), start with an organization's basic ability to install a standardized project management methodology for single projects covering pre-project, project and the post-project phases of the traditional project management process. In order of increasing complexity, we see the normal evolution of project management within an organization as it progresses from beginning to managing multiple projects to eventually grouping projects (either thematically in "buckets," divisionally based on the organization's business structure or perhaps strategically) into programs. Sometimes this results in a change in nomenclature and the organization starts talking about "program management" rather than project management. But the concepts are essentially the same and the only difference may be the concurrent introduction of a project management office (PMO) to support these efforts. This will be discussed in the next chapter.

If you consider the diagram opposite, you will note that having a project management office is the final phase of development of capability in terms of project level practices. Beyond this level you move into more complex, but much more beneficial, practices that involve working strategically at the enterprise level and across the entire organization.

Note also that there are two things moving in harmony in this diagram: as the complexity of tasks increases, so too does the realization of business benefits, as shown on the right-hand side. However, the accountability for the proper execution of these capabilities can be at the project level or at the organization level, as shown on the left-hand side of the diagram. Generally, most maturity models for project management published today support a similar hierarchy. However, I am not aware of many that culminate in strategy management—yet this is an essential point. Unless your project management efforts flow directly from a connection to strategy, they are bound to produce a less-than-optimal return.

Your organization's PPM process then becomes the central conduit for truly strategic projects. You'll recall that the objectives of the PPM process are project evaluation, selection and prioritization. Yet it is clear from our research that most organizations, in both the private and public sectors, have underdeveloped PPM processes, treating them as simple extensions of program management and centralized PMO capability. This is a mistake, given the potential impact of a truly strategic PPM process design. As a result, it should be treated as a highly strategic component within an organization's standardized project management processes and practices, and be carefully monitored by both project management professionals and executives for its effective functioning.

MANAGING PROJECT RISK

The remaining challenges are to focus on balancing risk and return against the available resources to optimize return (measured by realized strategic benefits).

Modern portfolio theory suggests that higher levels of return accrue to those who risk more and this concept applies to project selection as well. Obviously, risk is a natural part of the assessment of a project's "doability"; an organization must consider what the potential risk factors are that would not allow the project to come to fruition. In terms of project selection, this means that even if a project is highly strategic with high potential returns, if the organization faces challenges or known risks in its execution then the expected returns need to be adjusted downward to reflect that. This is known in portfolio theory as *risk-adjusted returns* and it is a critical concept to integrate into your understanding of effective PPM practices.

Risks come in two forms, anticipated and unanticipated, while the risks themselves can be internal or external and controllable or uncontrollable. These are standard labels well understood by those trying to assess business risks.

Traditional project management practice requires risk plans that anticipate typical risk factors and develop plans to mitigate these risks. Among my clients, I most often see risk factors that are external in nature: events which, if they occur, are beyond the control of the organization but which require a response in order for the project not to suffer a fatal risk. For instance, a natural disaster that wipes out your organization's data center is clearly something you would want to plan for, perhaps with a redundant back-up or outsourced solution provider that kicks in only when your center fails. However, this is hardly a risk that is unique to your organization. Anyone could face this, as it is a clear and present external risk.

When we consider risk this way, most organizations report they are quite comfortable with "risk-return balancing" and they make it a normal part of good project management practice. But if this is the way you assess project risk, it wouldn't vary much when implemented within a PPM context, as you would essentially continue to evaluate and price risk as you do today. So we are not going to spend significant time in this book on how to risk-adjust project returns for obvious external risks such as these. Any good executive team is likely in command of the risks associated with these kinds of external factors.

INTERNAL PROJECT RISK ASSESSMENT

On the other hand, our research has identified numerous internal risk factors that interfere with project success, which do not vary from project to project but which can be established at the organization level. I refer to these as internally dependent project execution risks.

Once I understood this, I began to think of project risk in a new light. Obviously you have to assess and prepare response plans for external risks; but my clients' projects were more often affected by internal execution risks than external risks. In fact, the approach is so powerful that the essential success of a PPM implementation depends on understanding and evaluating these risk factors, whose composition is unique to each organization. And since there is no way to predict them in advance or standardize these risks around a set of "common factors," it takes hard work to find, study, validate and document them for your particular organization.

So how do you discover what factors create complexity and risk in your organization that can potentially impair project delivery? The answer can be found in a research technique called *factor analysis*. The concept is fairly easy to understand, and you can find additional resources to help you with applying this in your own organization; we will cover just the basics here.

Risk factors for inclusion in your risk-scoring model are developed in a similar way as those you use for assessing strategic return: they are unique to your own organization. Perhaps they were part of past project failures (based on post-project reviews conducted after the fact). You initially get at these factors through qualitative research to identify what risks cause your projects to fail.

I recommend to executives and project managers that they go back two or three years and identify all projects which either failed or did not realize their forecasted business benefits. These are project candidates that can help us to discriminate specific, recurring risk factors. Once these are isolated, you can contrive standard definitions and attach values to assess internal risk factors.

Again, my approach is to use a simple scoring model. This may help, especially if you do not have an executive leadership team that is used to thinking about risk factors in a project-specific way. I usually build them as simple Excel spreadsheets.

But since risk factors are unique to each organization, these can only act as examples. Nonetheless, I have included a few typical risk factors used by some of my clients so you can see how this works. None of these factors should be adopted as standard risks, because the actual factors that drive project risk for you are specific to your individual organization. That is the power of this approach and what makes it so useful in assessing individual project risk profiles. This context-sensitive approach to enterprise risk management is significantly different in method from traditional models that only weigh generic or standardized risk factors for their likely occurrence.

One client organization has identified five specific factors which have historically caused its projects to either stumble or fail. They assigned standardized definitions of risk levels for each factor and attached a score from 1 to 5 to each definition. This way, internal risk is fairly treated for all projects equally. With this self-awareness, they are now able to score projects on the basis of the absence or presence of these individual factors, giving them a baseline risk level on which to adjust returns for any proposed project. The more present these factors are on a project, the higher its score—and the lower the project will be ranked for selection. If

Project	Number of Business Units Involved	Number of Resources Required	Technology Innovation	Scarcity of Key Resources	Inter-Dependen-cies	Risk Score
Project A	1	5	5	1	3	15
Project B	1	3	3	3	1	11
Project C	5	5	3	1	1	15
Project D	3	1	3	1	1	9
Project E	3	1	1	3	5	13

SCORE	Number of Business Units Involved	Number of FTE Required for Comple-tion	Technology Innovation	Scarcity of Key Resources	Project Inter-Dependen-cies	
(5)	>5	>30	Involves New Core Systems	Scarce Externally and Internally	>5 Other Projects	
(3)	>1 <= 4	15–29	Involves Major Changes	Scarce Internally but Contrac-table	= 2 to 5 Other Proj-ects	
(1)	= 1	1–14	Involves Minor Changes	Limited Scarcity/ Vendor Provided	Not Inter-dependent	

two projects offer similar returns but one has fewer internal risk factors, it will have an advantage. Using internal risk factors in addition to external ones, you create a subtle but important adjustment that helps accomplish a truly risk-adjusted portfolio selection at the total enterprise level as well as at the individual project level. Regardless of the risk assessment internal to each project, the executive team must assess the *general* risk levels of all projects. This score is on the absence/presence of specific risk factors, unique to the enterprise, affecting the overall risk of the entire portfolio.

Once this distinction is understood (what I refer to as structural versus technical or specific project risk), it becomes clear that each proposed project can then be looked at for its individual contribution to strategy and for its individual technical and structural risks in relation to the potential return. While traditional scoring models are less sophisticated and simply compare financial return to external risk, this approach proposes a scoring model that compares strategic contribution to total project risk, both internally and externally. This outcome while more complex, is more complete, and more dynamic.

The importance of this change in practice cannot be underestimated for its possible contribution to the private sector; but it is even more significant in the public sector. As previously discussed, the complex multi-stakeholder strategy present in the public sector makes traditional, financially driven project scoring models quite limited in their applicability. To address this imbalance, we must use a method that allows us to measure, for example, outcomes related to the greater social good or achievement of social policy objectives, not just financial terms (e.g., improvements in overall health levels, decreases in waiting lists or treatment delays, increases in success rates on standardized achievement tests, access to cost-effective day care or increases in the number of commercial patents issued to businesses). The BSC or any similar balanced strategy tool allows us to do that.

Furthermore, once strategic contribution is understood, we can use it to balance strategic outcomes with the costs of program implementation, enforcement, risk and other factors. This balanced view of strategy is imperative when responsibilities cannot be abandoned simply because they are costly. However, financial efficiency is still equally valued, thus creating the right balance for application of the methodology in this sector.

So we can account for the true value of a project—beyond its pure financial impact. And we can score the relative strategic value of any project over any other, by measuring its potential contribution to the overall performance measures.

Of course, since I earlier proposed that being "more strategic" in executive terms often means increasing our focus on strategy execution, there is another inherent value in this proposed change, and it is critically important. The organization needs to make sure it has enough projects in each dimension to accomplish its goals. This transfers the notion of "balance" from the scorecard methodology into a project setting and demands that the project portfolio also be similarly balanced. This helps build expected performance levels into the selection of the project portfolio from the outset. This is not the case with traditional, financially oriented PPM scoring models.

For instance, suppose you are considering a "greening" of your company that is going to both help the environment and also support your brand, because your customers will feel more comfortable buying your products; but the associated R&D project to create the new manufacturing process is going to cost a lot of money. Perhaps, because you are unable to fully quantify the benefits of "customer respect," and since increases in sales are forecast over many years, the project might show an initially low ROI. A financially oriented scoring model would give the project a failing mark. Prudently, while the project still requires close scrutiny, under our PPM model you would try to determine which other critical strategic measures (i.e., customer retention; margins; employee pride; media recognition) the project might positively impact, and use those as a strategic rationale to rank the project as providing a high return.

If your current scoring model involves measuring the relative financial performance of one project versus another, the only assurance a chief executive has is that the final approved project list is likely financially efficient. Yet what about its ability to achieve the overall corporate strategy? With the methodology proposed in this book, it becomes possible to consistently and practically measure potential strategy achievement of any project *in addition* to its financial efficiency. This ensures that we can "lock in" the organization's results by identifying projects that, while perhaps less financially efficient, are actually critical to some other dimension of strategy execution. This helps mitigate the risk of an imbalance towards short-term financial gains at the expense of long-term strategic capability and capacity building.

Significantly, leaders can now look across the project portfolio from this enhanced strategic vantage point and more easily determine if sufficient projects are available to actually achieve their organization's total strategic agenda. This is an important element of the proposed PPM framework that executives in the case studies really supported as being valuable (as you will see in Chapters 8 and 9).

This approach also finally enables the executive team to focus the organization's resources on those projects having the highest potential "on-strategy" contribution. You can eliminate projects earlier and reduce the amount of time wasted on estimating and planning projects that have no hope of ever getting approved. For one of our clients, this effort alone provided a 23% increase in available IT resources by reducing the number of projects being proposed that required effort estimates. The reinvestment of this effort into strategic projects also increased the speed with which the strategic projects were delivered and lifted the overall returns to previously unattained levels. This is the kind of benefit that awaits organizations prepared to make sustained, early calls on whether or not a project is actually strategic.

RESOURCE MANAGEMENT APPROACHES IN PPM

Any good PPM process must be initially designed to meet the three traditional gaps in project management practice that Dye & Pennypacker (1999) state as the objectives of a "best practices compliant PPM process":

1. Maximizing the value of the total portfolio
2. Balancing the portfolio against available resources
3. Linking projects with strategy

The parts of the process map we have discussed so far cover items 1 and 3 in this list quite effectively. (That is, one can link individual project outcomes to strategy through the use of balanced measures; and we have even added a significant advance in terms of risk assessment by establishing the notion of internal risk differing from external risk, to make these returns risk-adjusted.) This approach will therefore not only link projects to strategy but create an optimal portfolio of project selections that maximizes the value of all project efforts.

So next we turn to item 2 in the list above—resource balancing. The relevant steps are circled below. They require identifying available resources, and, as always, the steps look easier to accomplish on paper than they are in practice. Most of our client organizations have some difficulty in even accurately determining the resources available. Obviously, how to accomplish measuring available resources for projects is another step that is highly individual to the scale and scope of your organization's resources. This is another example of a challenging process that is worth the pursuit.

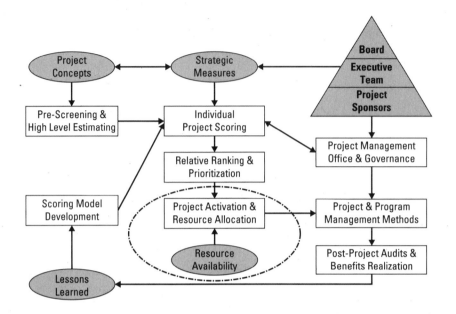

Resource Balancing Steps

The best project managers I have had the pleasure to work with during my career exhibit a clear and intuitive understanding of the importance of resource management and continually struggle with its impact on project progress. They are like chefs in a busy kitchen who have to make basic ingredients stretch to cover a multitude of orders during the course of the evening rush, all the while hoping they don't run out of any ingredients! This can be both distressing (for its impact on project results) and stressing (for its impact on them and their need to constantly balance and rebalance resource assignments within the portfolio).

If projects are properly assessed for risk and return, and are recommended for inclusion in the portfolio, we should then have the ability to prioritize these projects and absolutely make a case for all of them to be executed. This process generates an "approved list" of projects which is essentially nothing more than the complete list of projects in the portfolio. Only projects on this approved list should be worked on, although we have found that this discipline is hard to master, and some effort on phantom or unreported/unapproved projects still occurs. But this depletes precious resources, so learn to focus exclusively on your approved project list if you want to be most effective.

In more traditional PPM methodologies, we often find an interesting practice. Once the list of approved projects is produced (regardless of how the projects selected were scored), the list is often compared to the available resource base and a "waterline" is drawn to indicate projects that are "affordable" (based on capacity, available capital or some other similar criteria). This practice may appear to some executives as obvious or intuitive and it is quite common. But it is only "pseudo-strategic" rather than strategic. Why?

Well, if you have chosen these projects on the basis of a balanced strategic scoring model, then I am going to assume that each and every one is both valid and important to the execution of that strategy. Otherwise, they would not have ended up on the list, right? They should have been weeded out along the way and eliminated.

A sound PPM process actually treats resource availability as information required to determine when we activate a project—not whether we select it. This is a critical distinction that the generic process map clearly shows. Methods such as a master project schedule or similar tools are useful to establish when you should do a project, not if. But to summarize, if you pick a project as being essential to your strategy, your job is to find the resources to make it happen, not to rule the project out because the resources are not currently available!

INTEGRATING PPM WITH OTHER CORE BUSINESS PROCESSES

There is one final point to consider in the design of an effective PPM process within your organization. While we have provided you with a generic PPM process, it does not show the inter-relationships between the PPM process and the other core processes that most organizations have (for instance, a capital budget planning and allocation process or employee reward and recognition programs).

For example, one of the factors that makes a difference in terms of calculating available resources is just exactly how productive your workforce is. There are many who believe (and I concur) that some elements of productivity come from aligning rewards and recognition to the things you want employees to do. Or perhaps you might use bonus or incentive compensation to reward the kinds of behaviors that get projects done. If you are currently working in a private sector context, these kinds of tools are easily available to you and can be directed towards the strategic project outcomes you are seeking.

Unfortunately, this same approach is more difficult for the public sector to consider and implement. This is because of the natural occurrence of tenure-based compensation and high levels of unionization, resulting in position-based approaches to compensation. So, while it is more difficult to create this alignment, I must still challenge public sector leadership to understand the importance of these findings so as not to exclude their possible impact on organization project performance.

Still, most practitioners report thinking of PPM as relating to the selection and management of strategic projects. However, proper and effective portfolio management must also focus attention on the importance of aligning organization systems towards the successful execution of the portfolio once it is selected. In fact, it might be best to express the inherent concepts as a star diagram visually depicting what I mean. I call this a "systems circle" because for the middle to work, all the outlying drivers must also be aligned to the purpose of the center.

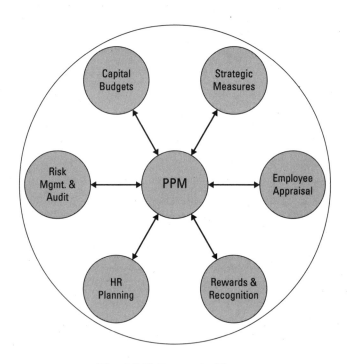

The PPM Process Cluster

Consider this: if we don't align strategic measures to the criteria we use to appraise employee contribution, we have a disconnect that will impact the execution of the portfolio. If rewards and recognition are not in place to correspond to what's important (the portfolio), then we are sending an inconsistent message to employees. We should be carefully looking at tenure, promotion and succession planning systems in HR to ensure they take into account the direct and successful execution of the portfolio as a major criterion. The same is true for systems such as risk management, audit and capital budgeting processes in Finance. This means that a truly effective PPM methodology cuts across the organization, drawing on supporting processes elsewhere with the intent to align effort to desired result—the successful selection and execution of the portfolio.

If your organization narrowly interprets PPM as being only about project selection, you have not made your full contribution as a project management professional to supporting your executive team's success. However, this means working in partnership with other functions in the business (HR, Finance, etc.) to ensure that internal systems align with the portfolio objectives, which in turn emanate from the strategic objectives of the firm. In this way, as PM professionals, we can contribute at multiple levels to aligning everyone in the organization to the strategy and improve organizational results.

So, once your PPM process is designed and implemented, you create an opportunity for your organization to realize more benefits sooner and to do so with a minimum of risk, while clearly stating how much resource will be required to succeed. Any C-suite executive is going to be pretty happy with this outcome and I hope that this chapter gives you the courage and support you need to forge ahead successfully with PPM in your organization.

## CHAPTER FIVE

Creating Small, Smart and Mighty PMOs to Steer the Way

Co-written with Lucy Proteau

Lucy Proteau is a project management consultant with more than 20 years' experience in leading large projects and creating lasting results for clients by instilling leading-edge PM practices, particularly implementing project management offices. In this chapter, she shares her knowledge of this important area with us and also guides the development of tools and methodologies on projectgurus.org to enhance project management practices everywhere.

When I first asked Lucy to contribute to this book, we went back through time and asked ourselves some key questions about the implementation of PMO's.

How many times over the last 25 years had she been asked to implement a PMO to support strategic projects? And how many of those were successful and why? Furthermore, in the consulting work that we have done together over the years, we were both convinced that often the project management practitioners and executive sponsors we dealt with were not getting maximum strategic benefit from their PMOs, and this frustrated us. A PMO can be a simple and powerful supporter of strategic work in an organization—but only if it's properly managed. We believe any organization, with some support, can improve its return on investment by steering the PMO to more effectively support strategic project selection and focused execution—and thus helping clear up the fog!

This chapter contains practical advice we've picked up along the way in an attempt to help clients select and deliver the right projects at the right time to support strategy execution. If you've never had a PMO, this chapter will help you kickstart your organization onto the path to success. And if you already have a PMO in place, you can use this as a simple checklist to determine if it's operating at a sufficiently strategic level to generate the outcomes you need to be successful.

Here are four essential elements of a PMO that we found directly helped to clear up some of the project fog that most organizations face.

- Get the PMO off to a good start by *defining its mission.*
- *Align the PMO* to what is strategically important to the organization.
- Help PMO leaders *understand critical relationship success factors.*
- *Build a culture for continuous improvement* for a maturing discipline.

Taking each of these key points in turn will help anyone associated with leading a PMO, and the key executive sponsors whose projects they are supporting, to understand how to work together to maximize strategic outcomes within their organizations.

GETTING OFF TO A GOOD START!

In some cases we have seen, it seems that PMOs are literally introduced into organizations "overnight," with very little fanfare or clarity around what they are intended to do or how they are going to do it. These people often add to the fog rather than helping to clear it away. In other cases, it seems that the PMOs are being set up as a last resort because all other efforts by leaders to get control of their project portfolios have failed. Rarely have we encountered a PMO that had a clear mission that directly linked project selection and delivery to what was important to the organization's strategy. Yet this is clearly the most beneficial role of a PMO: to provide focus and clarity around the execution of the most strategic projects in the organization.

Regardless of why your PMO was or is being introduced, remember that any lack of strategic clarity will eventually translate into gaps in execution that eventually undermine what you are trying to achieve. Here are some of the gaps we saw during the research for this book:

- Those involved in execution are unable to translate strategic priorities into appropriate action items or to make informed strategic trade-offs.
- Project managers quickly get frustrated when they work hard but the resulting outcome is deemed to be either a failure or only a mild success because, even though it achieved its intended result, the result was not ultimately very important.

- In the absence of clarity about the strategy, employees can't embrace their organization's highest priorities. Instead they get off track and focus on what *they* think is important rather than what *the organization* needs them to be working on.
- The lack of clarity impedes collaboration and synergy, meaning project teams can't build win-win solutions because it is not clear what they should be collaborating on, with whom, or why.

If employees are foggy about an organization's strategy, this fogginess will also surround the PMO's purpose. So people will simply make up their own minds about what the PMO should and should not do—and these perceptions can undermine everything you are trying to achieve. From one of our larger clients, we were amazed to see exactly such a response from a frontline project manager:

> So now that the PMO is here, they will somehow figure out how to fix everything in the organization that is broken about project selection, management and project execution processes so we can focus on getting real work done!

However, assuming this person's reality is properly reflected in this quote, there is little that the presence of a PMO can do to resolve this problem, and whatever the PMO does do will not meet expectations. The PMO will likely not become a value-added contributor to this project manager's work. In fact, a PMO introduced into this scenario will probably only make matters worse by adding another layer of confusion and obfuscation to already confused project management efforts. The problem of a lack of strategic direction, established already in the previous chapters of this book, is a key management responsibility and a PMO cannot operate in a strategic vacuum. Only courageous leaders who are willing to be accountable for charting the course and aligning the mission of a PMO to a clear organizational strategy can fix this problem.

To help correct this strategic deficit that we so often see, we suggest treating the introduction of a PMO with the same care as you would give to the launch of a new line of business, a new product or service offering, or any other strategic initiative. Recognize that introducing a PMO means change and that those involved in the process (including the leadership team) will need to approach project selection, management and execution differently if the PMO is to succeed in its mission. The effort to make the change happen will depend very much on your culture, your internal project management maturity and capability, and the will of the leadership team to make the change successful.

When we work with clients, we use a *very simple* tool we call our "Quick Step" method to help executives chart the course for change in their organizations. Many of our clients have found it works well to help them plan business changes. It involves six relatively simple and clear steps that are followed sequentially. Of course, the most important input to all of this is the organization's strategy and its strategic measures. They act as the input that determines what changes must be accomplished to support the execution of the strategy in the first place.

For this model to work, change must be described in *outcome terms*. That is, what specifically are we expecting will be different after we implement the change? For employees to get engaged in making the change happen, they have to see the rationale for the change clearly.

Based on our past work, we too often see organizations attempting changes that are either unnecessary or suspect—driven by personal agendas, preferences or perhaps good intentions but not by measurable strategic results that every employee can understand and support. This state makes change more difficult to accomplish.

The six sequential steps of the model are pretty straightforward and shouldn't require too much explanation to understand. On our website at www.projectgurus.org we have provided some practical tools related to using this model in a business setting (such as stakeholder maps and sample engagement strategies) that you may find helpful if you want to put this model into practice within your own context.

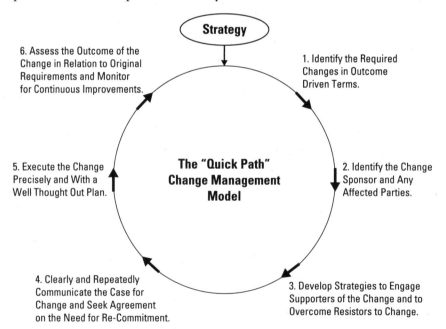

Strategy

6. Assess the Outcome of the Change in Relation to Original Requirements and Monitor for Continuous Improvements.

1. Identify the Required Changes in Outcome Driven Terms.

5. Execute the Change Precisely and With a Well Thought Out Plan.

The "Quick Path" Change Management Model

2. Identify the Change Sponsor and Any Affected Parties.

4. Clearly and Repeatedly Communicate the Case for Change and Seek Agreement on the Need for Re-Commitment.

3. Develop Strategies to Engage Supporters of the Change and to Overcome Resistors to Change.

Regardless of where you are in your PMO evolution, it's likely that changes in its form and function will be required over time. As you refine its mission, you can use this model to implement these changes and gain traction on moving forward quickly and succinctly.

If you have a PMO in place, take the time to solidify its mission, strategic purpose, structure, roles and responsibilities carefully to establish critical success factors for it to become a value-adding contributor to strategy execution. Then seek buy-in from a critical mass within the organization by communicating the case for change in outcome terms before moving forward with implementation of the change. Critical mass includes those people in your organization who are really needed to ensure success—and these individuals may or may not be at the top of your organization. Of course, it's also critical that your project managers buy into and understand the purpose of the PMO. (We often see executives trying to impose a PMO to deal with what they see as a project management performance problem that would be better addressed through constructive dialog and creating a shared purpose and mission between project managers and the PMO.) To help you in this pursuit, our next section will deal with how to get this dialogue going.

DEFINING OR RE-DEFINING THE PURPOSE OF THE PMO

An effective PMO influences strategy execution. So when you consider the role of a PMO within your organization, remember the following:

- It cannot succeed by dictating, controlling or simply tracking project status. Project teams across the organization must see the PMO as an ally and a resource they can come to when they need help removing real project roadblocks.
- The PMO will need to provide the right tools and practices to let project teams hit the ground running. Standardized methodologies are an important component of this; however, they must evolve naturally as practitioners see the value of standardization, re-use, common language, and so on.
- The PMO will need to walk before it runs and strive for ongoing continuous improvement. It must continually keep abreast of what is happening in the field of project management: this is a critical aspect of becoming a center of excellence within the organization for advanced project management practices.
- The PMO must learn to effectively coach and mentor for results using positive and appropriate methods that help professional

project managers feel good about addressing their own developmental needs.

To help you better define the role of the PMO in terms of outcomes, here is a framework that we have found useful and that we believe can eliminate at least some of the fog surrounding the role of a PMO within an organization.

Practical Tips for Defining Your PMO Purpose

- **Mission:** Why does the PMO exist? Your PMO mission statement must be simple and clear and not wordy. Make sure that anyone who reads it can figure out what you do!

- **Strategic Purpose:** How will it deliver its mission and purpose? What does the PMO have to do to align to organizational strategy and ensure that it stays on track to assist with strategy execution?

- **Key Success Factors:** How will we measure the PMO against strategic outcomes that matter to the entire organization and ensure that the investments made offer the organization a return?

- **Roles & Responsibilities:** Who is going to do what and who will the PMO report to? How will this reporting relationship provide the PMO with strategic influence in the C-Suite?

The purpose of the PMO can, and should, vary based on the needs of your organization, the stability or instability of its strategy, and the degree to which is has a centralized or decentralized structure. It will be affected by your organization's strategic focus, geographic scope, capability, competencies and executive preferences for how to implement the PMO effectively. All of this could loosely be summarized as being about your organization's culture.

We find "culture" to be an overused word in the world of business today. Everyone seems to be talking about, working on, complaining about or charting the risks and rewards of culture. However, what is missing for the most part is a dialogue about how we make culture fit the organization's strategy. Once again, the idea that there is a "best culture" is a myth. Only in the specific—when a type of culture is crafted to fit a clear strategy—can there be optimal performance.

Consider the following questions:

What are we doing and how does it affect me?
What's in it for me if we succeed?

Do we have the right people doing the right things all the time? What new capabilities will give us the capacity to succeed?

We often put these questions into simple pictures, like the one shown here, to help shape clients' thinking on this important topic.

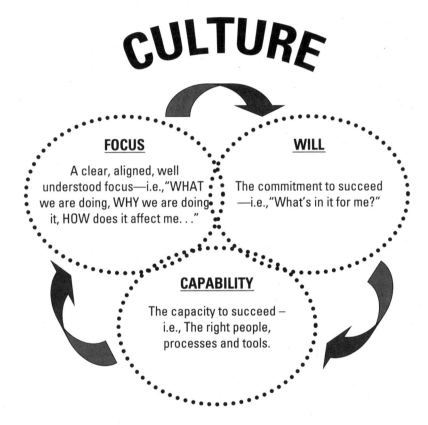

Often, we use the following mnemonic as a way to remember important aspects of the culture and its impact on strategy.

Let's consider this for a moment. If you understand clearly what you are dealing with at the outset, you will be able to quickly move from PMO implementation to adoption to maturity and a true focus on strategic execution. An understanding of culture enables you to bring people along with you as a leader—and to engage and energize employees about the prospect of being successful and achieving results.

For example, if your culture is very siloed today, with functions operating autonomously and without much co-ordination, you may find a lot

C	Commitment	How committed is the culture to the strategy and its execution?
U	Understanding	Does everyone understand the strategy and its measures of success?
L	Leverage	How well are we leveraging culture to successfully execute our strategy?
T	Tools	Are all the tools we need to support strategic success in place?
U	Urgency	Is there a shared sense of urgency about executing the strategy?
R	Responsiveness	Can we respond quickly enough to strategic threats and opportunities?
E	Engagement	What can we do as leaders to engage and energize the organization?

of resistance to moving from a culture where everyone essentially "does their own thing" to one of collaboration for selection of the right projects. On the other hand, if you have a hero culture, you may have difficulty building the structured discipline needed to do things in an orderly and proactive manner rather than on superhuman effort engaged at the last minute to rescue the project results. Heroes thrive on chaos, because they need it in order to charge in to save the day; they are often afraid that if this need no longer exists, maybe the organization will no longer have a need for them. Make sure when plotting how the PMO will impact your organization that you take into account your current capabilities, capacity, and what's in it for those affected by the change. Ask yourself a simple question: if you were being asked to change in the circumstances, how would you feel and what would you do about it?

By combining our model for quick path change with a clear understanding of your organization's culture, you can move insightfully and quickly to install a PMO that will be effective and create a fast-track process to support a renewed focus on the selection and execution of strategic projects. Or if you already have a PMO, use this model to assess how well it is dealing with the cultural and process impediments it may face before it can become a full strategic participant within the organization. If you do not have a PMO currently, use these models to determine if a PMO can add enough significant value to justify its investment costs, and then build only a small and mighty one!

STRATEGIC PMO PROCESSES

While there is certainly no "one size fits all solution" to your PMO needs, there are a number of functions and processes that recur among our clients who have achieved successful integration of the PMO into their companies. Here is a list of possible PMO service offerings that can help you get started if you are new at this.

Of course, this generic list of PMO services must align to the mission and strategy of the PMO, which, in turn, is aligned to the organization's mission and strategy. This will help you solidify your PMO's functioning within the organization, making it a small but mighty value-adding contributor. So, for instance, if you decide one aspect of your PMO's mission is contributing to project selection in support of strategy execution, defining this at the outset as one of its mandates will help ensure that this is clear to the entire organization.

Taking the time to determine how the PMO will work with its customers to deliver its purpose is another essential ingredient in PMO success. Questions to ask around this point include the following:

- Who does the PMO work for?
- Who are the PMO's customers?
- How does the PMO get buy-in from PMO customers to approved methods for project selection?
- How will we manage to produce consistent execution across the organization?
- How can we make sure there is enough capacity to deliver the "lights-on" and mandatory or regulatory type projects concurrently with strategic change projects?

If your PMO is brand new, review the implementation strategy for the PMO discussed earlier and decide whether you want to go forward with a "big-bang" approach or a "phased-in" approach. Again, the answer to this question depends a lot on your culture. What is your starting point for acceptance of the idea of a PMO, and is it likely to be well received? We call this "change dynamics" and it involves assessing the extent to which a change will be viewed as major or minor in nature. Obviously, this affects how much you have to manage the change process within the organization around the introduction of the new PMO.

There is greater acceptance of things that have a proven track record of success; this suggests that in very large organizations you go with a "phased-in" approach where possible. This gives you time to pilot-test

Representative PMO Activities

LIST OF POSSIBLE PMO SERVICE OFFERINGS (Alphabetical Order)	Yes/No
Benefits Tracking	
Change Management & Change Control	
Communication—includes PMO marketing	
Contingency Planning	
Cost Estimation/Tracking/Management	
Cross Business Function Project Management (Matrixed or Direct?)	
Customer Service	
Planning	
PMO Tool Support	
Problem Management	
Process Management	
Project Governance	
Project Methodology (Scaled to project size/risk)	
Project Portfolio Management	
Project Portfolio Oversight	
Project Selection in Support of Business Strategy Execution	
Project Skills Upgrading	
Project Toolkit for Reuse	
Provisioning/Purchasing Policy & Practices	
Quality Assurance	
Reporting	
Resource Planning	
Risk Management	
Other. . .	

Source: www.projectgurus.org

your PMO service, offering it to a smaller but respected division of the company first. Once you have worked out the kinks, you can extend the service offering out to the entire organization.

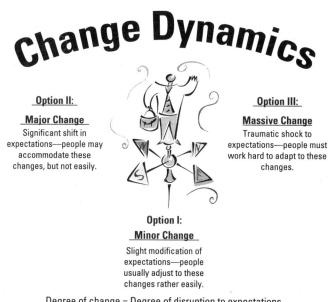

Change Dynamics

Option II:

Major Change

Significant shift in expectations—people may accommodate these changes, but not easily.

Option III:

Massive Change

Traumatic shock to expectations—people must work hard to adapt to these changes.

Option I:

Minor Change

Slight modification of expectations—people usually adjust to these changes rather easily.

Degree of change = Degree of disruption to expectations

KEY SUCCESS FACTORS

Now that you have established your purpose and identified your strategy for the PMO, you will need to determine how to measure its service offerings and levels to see whether or not it is actually achieving its intended purpose. Since we know that the only things managed in most businesses are things that actually get measured, this is a pretty important step. Take the time now to:

- Establish the minimum standards of performance. Be sure to consider any unique internal policy, regulatory or other requirements.
- Measure how well the PMO is delivering against its purpose and the minimum standards; publish the results organization-wide.
- Let the organization know how it is doing and what is being done to continuously measure and improve the service offering.

All this requires that you find formal review methods for your PMO. This can involve using customer feedback mechanisms (surveys, ratings, publishing performance scores) and also negotiating specific service level agreements between the PMO and key customers, if your organization is large enough. Either way, a good PMO is constantly judging itself relative to its internal customer base!

After getting through everything else, most people would think that defining the process roles and responsibilities would be easy. Maybe. It all depends on your organization's culture and willingness to accept responsibility for project outcomes—and for outcomes, period.

To establish these all-important lines of responsibility, we use a tool called RACI (Responsible, Accountable, Consulted and Informed). This simple chart helps sort out who does what for any business process or project you have within your organization. We rely on this tool extensively in our practice to help clients shape accountabilities in specific terms. It's similarly useful as a tool for the PMO to define its own processes.

A Sample RACI Chart

PMO PROCESS ROLES & RESPONSIBILITIES	PMO	Project Requestor	Project Management Committee	Project Manager	Etc.
Planning & Project Initiation	R	A	R	R	
Project Selection	R	R	A	C, I	
Project Governance	A	R	R	R	
Project Methodology (Scaled to project size/risk)	A	R	R	R, C, I	
Project Toolkit for Reuse	A	I	I	R	
Other					

If you have used RACI or similar tools before, this chart is self-explanatory and the benefits of its use will be obvious to you. If this is a new tool, you should visit our website at www.projectgurus.org to download a detailed explanation of the chart and its use, and to see examples of how to build and implement a RACI chart for a simple business process.

RACI can also help address the issue of where a PMO should report or to whom. In some of our client organizations, PMOs report centrally and directly to senior management. In other instances, they may be divisional or functional. Regardless of the ultimate reporting structure, clear and strong support from senior management is necessary for a PMO to function effectively, and you can use a RACI chart to secure and maintain the required level of executive support you need to succeed.

SUPPORTING STRATEGIC PROJECT SELECTION

Once established, how can the PMO partner with the leadership team to help them select the projects that truly return the biggest overall benefit to the organization? This is the defining question of this chapter. One of the most important considerations, if you want a PMO to make a strategic contribution to strategy execution, is how to help your organization develop and implement its own PPM process framework. We counsel that your PMO drive the PPM process within your organization and we have, as a result of our work with firms around the world, consistently seen four benefits of this approach:

1. Improving selection of the optimal project portfolio (strategy alignment)
2. Optimizing projects, resources and total spending (project performance)
3. Managing the portfolio in one place (visibility; central co-ordination)
4. Getting up and running in weeks, not months (quick ROI)

Since many large projects can run for several years, how can we keep the focus on doing the right projects until they are done? And how do we ensure that the process of balancing and adjusting the project portfolio remains dynamic? Some of that can be seen using a simple process diagram that illustrates how an evergreen project submission and approval process can help with this question.

IMPROVING THE PROJECT SUBMISSION PROCESS

The whole idea behind trying to improve the project submission process is to ensure that your organization works on the right projects at the right time. As we have already seen, you need to establish clear criteria to help employees define what a project actually is and how to select the ones that are most important now. We do this using the kinds of scoring models we described in Chapter 3. But keep in mind that you will always need to deliver new strategic projects concurrent with "lights on" and mandatory

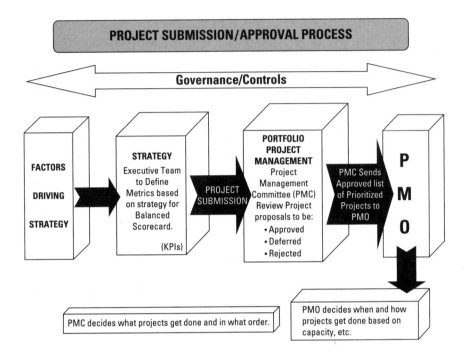

or regulatory type projects, and so your project selection process has to distinguish between projects you can and cannot choose to do.

Once we have established what a company's strategic goals are, the PMO can then help facilitate a process to select the projects that best achieve those goals. But to be useful, Key Performance Indicators (KPIs) must be quantifiable, agreed-upon beforehand, and reflect the critical success factors of your particular organization. For this reason they will differ for each company: no two organizations will ever have identical scoring models.

For example, a business may have as one of its Key Performance Indicators the percentage of its revenue that comes from products introduced within the last two years. This would indicate that management's focus is on new product development as a strategic priority. We know this because the measure must connect to the strategy.

However, even though a university is a business in many ways and it may focus on similar things, its KPIs will be expressed differently. Perhaps it might focus on the percentage increase year over year in student enrolment. While you can see the parallel intent (both focus on revenue from new sources and growth in market share), each measure is specific to the organization and its focus. As the PMO, you must be able to understand every one of your organization's key measures and ensure that you understand how they link to its strategy.

So what happens if you work in an organization that is not very strategic or where the measures used have not been clearly identified? In that case, it is a good idea to stop and go through the following exercise to at least give you some idea of what is important.

> *For our organization to be successful, we must be good at the following activities . . .*
>   1._____
>   2._____
>   3. _____

Once you know what's important, you are in a position to develop tools that can be used by everyone during project conceptualization. These tools can be used by internal teams to assess, score and submit the right projects—every time. Remember, it is essential that everyone buy into the PPM process before it is implemented to avoid some of the common project selection techniques we so often see in use:

- He who yells loudest wins.
- He who generates the most revenue (right now) wins.
- We have no choice—we should have acted sooner but . . . the guy yelling didn't want to hear about it.
- Nobody is sure . . . the project just popped up but everyone says it's really important.

Since the contents of the scoring model are going to determine which projects get selected, it is important that they be clearly available and understood within the organization and used consistently. If your organization is not using a balanced scorecard (or similar tool) today, it may not be common to publish performance results broadly. Yet this is essential information for those developing and submitting project proposals.

The following idea might be helpful in such a situation. It is a template that we developed with one of our airline clients, which they use as part of their internal project selection process. Think of it as "Scoring Model Light." This process was introduced and managed by the PMO as a gentle first step towards a more strategic use of project resources.

It provides guidance to those submitting projects about what exactly the business outcomes are that a project should target before it will be successfully selected to proceed. While it makes use of a simple set of scoring guidelines, it does not attempt as complicated a connection as defining the specific level of contribution of each project's deliverables to each measure on the scorecard or list of KPIs. Rather, it takes a simpler, holistic view of a project and the company's strategic goals to summarize strategic project impact.

While it cannot ensure a balanced portfolio (because the scoring model and tools do not call for that), it at least begins the process of breaking down why a project is or is not selected on the basis of measurable contribution; this is an excellent early "win" for a PMO within any organization newly embarking on portfolio management.

Regardless of the approach you choose to use, what is essential is for the PMO to facilitate establishing agreement within the organization on the consistent use of one method.

Something else we have learned about effective PMOs is that they work to actively support project sponsors through the completion of the project submission process. We have heard this characterized as being a "pair of hands" or an "administrative role," and some project managers we know resent our suggestion that they do this work. However, we feel that by doing it this way, the PMO also gains in the following ways:

- Gives the PMO insight about future projects
- Establishes a good relationship with the project sponsor early in the process, which is critical to overall project success
- Allows the PMO to ask some of the tough questions that are bound to come up at project selection time ahead of the submission completion.

ESTABLISHING STANDARD METHODOLOGY (OURS OR YOURS)

Once projects have been selected to proceed, employees begin to feel the real value of the PMO. A well-structured and organized PMO can not only help with standardizing and operating the project selection process but can ensure that projects use a standard project management process too. Why bother with this?

The primary answer is that the process facilitates portfolio reporting, risk assessment and management, and work related to benefits realization and post-project reviews. It assists with developing and implementing appropriate capacity planning and project activation and reporting tools. And it just makes sense.

Key Performance Indicators (KPIs)

Scoring guidelines are listed below:	Financial (OPS):	Customer & Markets:	Internal Process:	Learning Growth & Employees:
HIGH (50) = 2 or more KPIs directly impacted. **MEDIUM (25)** = 2 or more KPIs indirectly impacted or 1 KPI directly impacted. **LOW (15)** = Low impact to any of the KPIs. **NONE (0)** = No impact	1. Reduced Cost/Available Seat Mile (ASM). 2. Targeted Cost & Reductions (P&L Items) *Financial (CORP):* 1. Increased Profitability 2. Increased Return on Assets (ROA).	1. Improved Safety 2. Improved On-time Performance 3. Increased Passenger Satisfaction 4. Increased Tour Operator Satisfaction	1. Better Labor Efficiency—i.e., reduced overtime. 2. Irregular Operations (IROPS)—i.e., reduced cost/incident based on cost/delay minute @ $75/minute. 3. Fewer Reportable Events to Transport Canada (TC)	1. Lower Voluntary Turnover. 2. Higher Employee Satisfaction 3. Better Return on Training—$ spent per employee. 4. Higher % of employee exceeding performance targets.
TOTAL				

While project managers must have the flexibility to abbreviate or eliminate steps that do not align to their project size, approach, complexity and risk, at least knowing the right steps to take is an important contribution to internal project management knowledge. Deviation is generally done with the approval of the PMO to see that minimum work is done to ensure positive project outcome—as we all know, people love to take shortcuts. For many projects, the results of not being detailed enough can be disastrous. For example, badly defined work scope typically results in unhappy sponsors or project deliverables that do not meet business needs at all.

We often hear complaints such as, "We don't have enough time to define requirements, document the design of our solution, etc." If you don't have the time to do the basics needed for a good result, you'd be better off not doing the project until you have the time to do it right.

An effective PMO also recognizes that its internal methodologies need to take into account the organization's other functions and aspirations, and its unique culture or approach to certain things internally, such as software development methodologies, as one relevant example. These are strategic conversations through which the PMO can gain credibility by initiating, facilitating and co-owning outcomes before putting any changes into company-wide effect.

SUMMARY OF DO'S AND DON'TS OF AN EFFECTIVE PMO

- Do remember why you exist. Revisit your purpose each year to ensure it is still valid.
- Do meet with your customers regularly to get a sense of what is working and what isn't. This will help you find ways to continuously improve your service offering.
- Do support those involved in project execution—especially the project managers. They need to see value from the PMO and you need their support for your continued success.
- Do look at ways to make things easier for those doing the work. Encourage and facilitate knowledge and tool-sharing.
- Do encourage the execution teams to do the right amount of documentation to ensure that those who support or enhance solutions have enough information to do so effectively.
- Do facilitate the development of project management skills in the organization.

- Don't assume everyone will buy into what you are doing, even if they pay lip service to it.
- Don't be too rigid. Staying flexible will usually result in more than one way to get good results.
- Don't make things more complex than they need to be—keep it simple!
- Don't try to do everything all at once. Move incrementally and deliberately, always with your mission and strategy in mind.

## CHAPTER SIX

The Role of the Board: Integrating Measurement and Accountability for Project Results

Co-written with Janice Wooster.

Janice has spent more than 30 years working in the private, not-for-profit, government and education sectors in areas such as administration, human resources, training and development, customer loyalty, business process design and project management. Having finished her pre-consulting career as a vice president for a very large non-profit organization, she has seen first hand some of the challenges of dealing with boards and will outline her experiences and suggestions in this chapter. She also regularly contributes to the content of www.projectgurus.org; you can find her there if you have something to share.

Boards function differently in different organizations. They exhibit a range of interpretations of the role of director, and they all execute the important role of governance with a slightly different tone, tenure and effectiveness. While being a project manager is challenging at any time, with high-profile projects you are faced with the additional challenge of managing the expectations and involvement of your organization's board of directors. Much has been written on the topic of board governance. In my experience consulting to boards, very little of it has dealt with the role of the board in terms of *project* governance and management. With the increase in project-oriented work environments that we have seen, especially in the last decade, dealing with governance becomes an important leverage point for strategy execution.

It is something we must pay more careful attention to as project managers. You will recall the PPM process diagram from Chapter 1. It included clear references for board involvement in project governance, if your
organization's project portfolio management process is to be complete.
Therefore, the focus of this chapter will be on the upper right-hand quad-
rant, circled on the chart below.

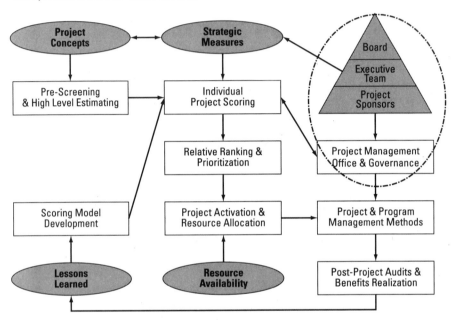

Whether you are a board member, senior executive or project management professional, you can quickly ascertain that your specific accountabilities in this process change. And it is not our suggestion in the diagram
that board members become directly involved in project management;
rather, we see their role as project governance and input into and approval
of the strategic measures shown by the arrows.

Note also that we see the board working through the executive team
and project sponsors, with little direct interface with the organization.
This is no different from the distinction we make in terms of clarifying the
role of executive leadership versus governance and stewardship. However,
what this diagram does suggest (and what our research supports) is that
we must *actively* generate a shared collective interest in maximizing the
selection and execution of the most effective project portfolio. Otherwise,
we risk that the governance role can move from supporting execution
of strategic projects to blocking or stalling internal project management
efforts.

But what are you to do if your board or executives do not support your involvement at this level of the strategic process? Understanding the "ins and outs" of a process is a great starting point. We always suggest beginning with an understanding of the role of governance so that you might be in a better position to help justify why, as a senior project manager, you should be involved in some aspects of this process.

Similarly, if you are a C-suite executive, what are you doing to promote the proper involvement of your board in the governance of key strategic projects? Have you considered what role a project manager might play in helping this outcome? As long-time consultants to industry, and often at the CEO/COO and board levels, we encourage our clients to determine what level of engagement is required among all the parties concerned (board, executives and PMs) to assure that the organization can achieve its strategic purpose.

Let's consider different types of boards and the ways they typically function.

ALL BOARDS ARE NOT CREATED EQUAL

Although every board is ultimately responsible for "governance," roles and specific responsibilities may be similarly described but vary significantly between the corporate and not-for-profit sectors. This happens with good reason. The scope and magnitude of responsibility of the board chair of a multinational corporation includes a level of complexity and responsibility that is unlikely to be present in the same position on the board of a community-based not-for-profit corporation. So, while each may be described as a "chair" and each is considered responsible for managing the board in its role as governors of the organization, the implementation and how they go about doing this won't be the same. As a result, a project manager needs to understand those differences and as part of project planning, consider how best to work with this unique, and often very powerful, stakeholder group.

Similarly, we often address ourselves to executives and boards while challenging them on the organizational environments they create. As you will see in Chapter 7 on leadership roles in projects, we have a particular perspective that culture within an organization must be deliberately and thoughtfully created. In the absence of this specific effort, culture develops on its own and often creates a strangulating effect on strategic performance by focusing the attention on the exceptions rather than on the rule. Over time, as governors and managers spot issues or concerns, they often

Decreasing Rates of Strategic Return for Time and Attention Invested

Higher Empowerment

What

Manage Process

Manage People

Who

Manage Tasks & "Things"

How

Lower Empowerment

The Performance Ladder

act to handle such concerns by creating a policy, procedure, or rule as a means to avoid them. While seemingly well-intentioned, these policies creep down the rungs of the "performance ladder."

The most effective boards and executives refrain from managing the specifics of things that need to be done. These are always at too low a level for them to correctly diagnose; trying to do so would take the board deeply into operations that are best left to frontline staff. Those are the people who know how to get things done within the organization.

Similarly, if boards and executives try and exclusively manage "the people" (often with the naïve notion that if we simply select great people we get great performance—a finding numerous researchers have proven is not true), they leave ultimate results up to the competency of individual leaders. In doing so, they create a culture of cowboys and heroes, focused on getting things done their way and not always to the benefit of the larger organization.

To actually get the balance right, boards and executive teams must focus on managing processes. This means understanding organizational capabilities and performance gaps, establishing measures and targets that challenge the organization to perform, and acting with purpose, ensuring processes can generate strategic results. These are all key to long-term success. Finding that balance is about creating a *context* for work efforts—both as leaders and as workers. But the who and how are less important to a board than what is actually accomplished.

This approach creates an empowered and engaged culture. We've found it allows people to use their smarts, dedication and experience in order to figure out how to create the solutions to the process problems that need fixing: this is the root of high performance. Further, the board and executive team's approach establishes the culture of the organization.

There is another important nuance to be aware of as a project manager who may aspire to work at the highest echelons of your organization: even while the boards of two large multinationals may look the same in many ways (e.g., number and type of directors, committee structure, roles and responsibilities), it is the desire of the collective to be a proactive and responsible board that will drive them to a natural curiosity about how they fit into project management within the organization. A more activist board, with a willingness and desire to govern more directly and accountably, will embrace that role and participate fully. You must nurture this interest carefully and help distinguish between active involvement in governance versus interfering in the actual management of projects. This is a delicate balancing act.

On the other hand, an unwilling or more tightly aligned board that is providing less governance and insight to management may not see a natural fit between themselves, their role and strategic portfolio management. (Here again, one size does not fit all; the boards in any of your organizations may operate differently.) Here your role is to gently champion the board's involvement and have them see the strategic contribution that a well-run PPM process can deliver. Either way, working with the board on any process is a journey of discovery for the board itself and for you as a project management professional.

But what can you do if you are not directly involved or the invitation is not accepted? Work through your project sponsors. Boards may not realize at first that there is a natural role for them to play in the PPM process. Perhaps they see it as too operational or managerial, and they may be unaware of the consequences of not actively managing the risks of the project portfolio at the highest organizational levels. But your sponsors do—or they should.

Ideally, as a project management professional, your relationship with your sponsor can provide an opportunity to influence this behavior inside your organization, by getting your CEO and the executive team to see the value of extending the PPM process to the board level. But ultimate success will come over time from an understanding of how ultimately strategic the PPM process is to any organization. This is a message we need to get out to everyone involved in the process. Selecting the portfolio to align with strategy is not simply a managerial task; it is an iterative and complex balancing act between risk and reward that ultimately determines competitiveness, effectiveness and efficiency. In our view, CEOs and boards should be involved in questions around these central points.

FOR-PROFIT OR NON-PROFIT BOARD: WHAT'S THE DIFFERENCE?

The general role of any corporate board can be summarized in three points:

1. The board is accountable to shareholders for ensuring that the company achieves appropriate returns (profits, return on equity, increasing the stock price to create capital gains)
2. The board must avoid what is unacceptable or anything that impairs returns (such as excessive risk, unethical or illegal conduct, unproductive investments or write-offs)
3. The board acts as a proxy for shareholders' interests and is particularly responsible for selecting and hiring the CEO, defining success and failure for him/her and conducting sufficient and appropriate performance management of the incumbent CEO to assure the appropriate execution of strategy that will create the outcomes in the first two points above.

What is most significant here is the focus of the corporate board acting as a proxy of the shareholder; a stakeholder group that does not exist in the not-for-profit world. The shareholder becomes the primary customer of the corporate board. A not-for-profit board may identify primary customers as donors, funding agencies or service recipients. Identifying the not-for-profit customer is a more complex (and potentially a politically challenging task) for a board that lacks the defining profit motive of the private sector.

Who "owns" the non-profit? Certain kinds of non-profits, such as trade associations or benevolent societies, are most often owned by their members. As for agencies providing social services, health or education, we can think of them as being owned by the communities or groups they serve. Regardless of how we define ownership, it is imperative that project managers and executives understand who the board is supposed to represent in its actions.

Once this starting point is known, and the non-profit board has established who it serves and why, much of the rest of the governance practices that apply in the private sector also apply in the not-for-profit sector. The board will normally establish clear performance criteria on which to assess organizational and CEO performance. (Incidentally, these measures are often quite different for the for-profit and not-for-profit sectors.) These measures help define how the organization will operate and what management will report on to the board. There may be regulatory components involved if the not-for-profit has some kind of reduced-tax or tax-exempt status in a particular jurisdiction. This also plays a part in the role of the board in any particular organization and defines for what it may be legally accountable. If you work in a regulated industry, it is also imperative you understand in detail the complexities of the regulatory framework for your industry.

Once the mission and strategy are established, they should be made clear throughout the organization, and any project manager or executive should have sufficient transparency on this point to assess how the board is structured to represent their stakeholders. In some instances, project managers may have to query management about the clarity of the mission or vision, especially if they are not clear. This must be done in a professional and encouraging way; we often counsel that the most appropriate path is simply to seek to have management articulate strategic goals in measurable terms. You might use questions such as "How can we be sure that any one particular project is more strategic than another?" or "Can you help me see the strategic connection between this project's outcomes and our strategy?" These kinds of questions will be enough to prompt the dialogue necessary to move forward and implement at least modest improvements in how projects can be strategically aligned and selected.

ELECTED OR APPOINTED?

Board members may either be appointed or elected. It is important to know not only how they arrive at the board but how they were selected and the perspective they may bring to their role based on past experiences. This can often be gleaned from board members' CVs. Usually, candidates for boards of directors are required to go through a rigorous screening process, and appointments are competitive. But in more informal settings like parent-teacher councils, community organizations or clubs, they may simply put a hand up and volunteer.

While board members are often selected because they bring a specific expertise to the table, they may also be recruited in order to access their fundraising or political network, or, in the case of community service boards, they may be representatives of people who are part of the service delivery or client networks. In for-profit boards, they may arrive as a colleague or acquaintance of a current executive or perhaps be specifically recruited to be part of a "slate of directors" that is proposed and voted on at a corporation's annual general meeting.

Remember, for many board members, the experience is something new. Effective organizations have a board governance model that informs the members and officers of their responsibilities and roles, and draws a clear line between governance and operations. This is usually sufficient for standard operations; however, high-profile special projects (such as the wholesale change or updating of critical systems) that do not fit into the operational model may garner more board involvement than is appropriate. This happens most often in organizations where project risk is not really understood; board members equate the size of a project with risk, when this may not be accurate.

So this is a really crucial point in this chapter: in order to avoid the pitfalls of either always guessing which projects to present, or presenting them all (and, in our experience, neither of these is satisfactory), it is critical that your organization's board governance model give clear definitions of projects that require board approval. Explain which projects need board approval in order to proceed; describe, also, which projects must include regular progress reports, because of their high risk profile or importance to strategy execution.

Working closely with your executive team is another critical area where the expertise of project managers can help define the internal "better practice" you wish to follow. For example, one of our major companies (an international insurance company) requires that any project that has the ability to affect the P&L by more than 5% either way be tracked and monitored separately by the board. This is an interesting and simple way of defining projects that require scrutiny. They concluded that 5% variance from forecast creates sufficient risk or opportunity to involve the board.

Another one of our clients (a large central government agency) is subject to more traditional thresholds of spending; any project that involves total costs of more than $1,000,000 requires board approval and oversight.

Of course, for our clients, we most often recommend using the risk scoring models we outlined in Chapter 4 as a way for the board to determine which kinds of projects pose risk and then set a threshold for regular board monitoring.

Obviously, involving the board in these ways implies the potential for project cancellation. This is easy to forget. In our experience, it's a challenging issue in smaller not-for-profits where boards play a more hands-on role. It is not a good use of the project manager's time to try to move them to a more hands-off position, because smaller organizations may be intentionally set up to utilize board expertise in operations. The benefit of this is that it often provides access to specialized resources that would otherwise be unaffordable.

Regardless of the degree of involvement by your board, become engaged in leading discussions about these important topics. Become familiar with the organization and operation of your organization's board; become an advocate for including the PPM process; and support your executive team in that engagement for optimal strategic results. Doing this will distinguish you as an unusual practitioner who understands the proper place of projects as the building blocks of strategy.

BOARD COMMITTEES

Most boards have standing committees in major operational areas such as finance and compensation. Often these involve areas of high risk for compliance with regulatory frameworks, such as the requirement of the *Sarbanes-Oxley Act* in the U.S. around financial statement accuracy or executive compensation disclosure requirements of major stock exchanges. These have become increasingly common in jurisdictions throughout the world, and individual board members may face personal liability for corporate failures. Board members are keenly interested in any project that touches these important areas. While these committees are essential to board operations and the fulfillment of good governance, they may test the boundaries of the role of a governor versus a manager. As a project manager who may be in front of a project in one of these areas, this could involve reporting to a board committee as well as day-to-day operational reporting to your project sponsor on the project's terms of reference and its progress. This can create unique challenges.

In this instance, beyond the overall board governance model, it is now important for the project manager to also consider the mandate of the committee and its composition, and how that can translate into a revised process for project reporting. Setting boundaries around how much information will be reported to the committee and in what way, how they will be involved in the decision-making process, and what decisions require their approval is critical. Similarly, active discussions with your sponsor about how you will both interact positively to ensure accurate disclosure to the board is essential. Do not be surprised if your executive sponsor is nervous about your involvement in board reporting; this is a normal reaction. (You may also find the experience anxiety-provoking: the stakes seem so high.)

In our experience, this happens for two reasons. First, the level of collegiality between the executive team and the project manager is usually greater than between the project manager and the board. That more casual familiarity may cause the project manager to drop some of the barriers between management and governance and to treat executive conversations less formally. This can inadvertently cause you to make assumptions about what the executive team does and does not really understand about your project and its progress.

Similarly, we caution executives to make sure they isolate their respect and appreciation for the project manager as a colleague from the necessity to be absolutely clear about project status at any time. This is not a matter of trust; it is about the respective roles of project managers and executives. While each may trust the other, it is important not to trust your assumptions about what is happening on the project.

Similarly, while project managers may want to have appropriate collegial relations with board members, it is important to recognize that the board are not experts in the business and do not have a direct operational role. That is the domain of executives; PMs may confuse this point. Boards are there to provide oversight and governance—they are not peers or supervisors to your executive team (or simply the next level up). This means that a certain amount of distance between project managers and the board is actually quite important. Managers appearing before a board should not deny them information; on the other hand, if you initially treat them as you do your executive team, you may find you have invited board members to cross a line that is difficult to redraw, now and for future projects. Project managers and the executives they serve need to establish patterns of governance of projects with the board, not the management of projects. As a project manager, you must actively work to make the distinctions clear, in terms of both the process and your own behavior as a professional.

Keep in mind, you may quickly find yourself as a confidant in, an influence on, or even a participant in, internal conflicts between committee members or between the board and the executive team—difficult and potentially dangerous territory. Take care about being drawn into communications with individual committee members, such as phone calls to ask for updates outside of the committee meeting schedule. In your desire to keep committee members appropriately informed and happy—and as a project manager, you already know how difficult this process is—you may at best establish a reporting structure separate from the committee, and, at worst, appear to align yourself with one side of a conflict. To the extent that you can, avoid informal or offline communication; your relationships with the board should be governed by a transparent and clearly understood protocol for communications that your executive is aware of and has approved.

THE BOARD AS PROJECT SPONSOR

In some high-profile or high-cost projects the board may be the project sponsor. Ensure that the definition of a project sponsor is as stringently applied to the board as it is within management. If the board is to take on the role of project sponsor, they must actively fulfil all aspects of this role as defined in the internal project management process.

In fact, the role of an effective project sponsor serves as a very good model in general to define the role of a board in relation to special projects. The project sponsor is the public face and voice of the project to his or her peer group, in this case the board, and supports the work of the project manager in these ways:

- Providing clarity regarding project objectives
- Approving allocation of the necessary resources
- Approving or gaining approval for changes in project scope, budget/resources or timing
- Securing broad commitment to the desired outcomes
- Ensuring appropriate escalation and quick resolution of issues affecting the project.

An effective sponsor understands the intricate network of relationships that makes up the organization, knows how the project will affect others in the organization, and assists in gaining buy-in from key stakeholders.

The real challenge of the board as project sponsor is that instead of one person being the sponsor, as is typical, you may have anywhere from eight to 15 people as your project sponsor. While each person may offer unique abilities and perspectives, those differences are hard to manage.

- Use an agreed-upon template for reporting to ensure that you remain focused on alignment with strategy and don't accidentally stray into operations.
- Agree on one board member to act as your point person to receive reports and answer questions from you or other board members.
- Quickly escalate breaches in protocol to the CEO or ED, who has ultimate responsibility for board reporting.

We cannot specifically provide consulting advice in this book because the situations you each need to address are all so different, but we have provided process suggestions to help you resolve the most common challenges you may face as a project management professional working between the executive team and the board. If this is an area of concern for you, you can become a member of our practitioners' community at www.projectgurus.org, where you can share your own experiences as we seek to collectively improve and share the better practices of professional project management globally.

THE BOARD AS A PROJECT RESOURCE

While increasing the involvement of the board in any project carries risks, there are some projects for which the rewards outweigh these risks and we should get board members more actively and directly involved.

I was once responsible for a multimillion dollar corporate head office relocation project that included the design, build and fit-up of a brand new facility and sale of an older building in another part of the city. Needless to say, this project was front and center with the organization's board due to its high visibility with the public, level of risk, cost and potential for business interruption.

Because members of this board all had relationships and access to networks that included highly expert and influential people, they proved to be a valuable resource when it came to identifying potential partners for real estate development, office furnishings, specialized subcontractors and relocation services. Their ability to open doors and create buy-in with potential suppliers saved time and dollars overall. However, the organization's decision-making protocol still had to be applied.

If your organization has a request for proposal (RFP) or request for information (RFI) process that seeks a minimum number of bids, then these must still be followed—even with a personal recommendation from a board member. Since board members may be people who are used to making independent decisions in their personal or business lives, they may struggle with their zeal to be a resource and the need to step back and avoid crossing that internal governance line. So you want them to lend their expertise and counsel as the primary mandate without actively making the decision itself. This requires that you, as the project sponsor, remind the board generally (as opposed to the individual in question directly) about the need for the project to conform to the organization's policies and procedures (and doubly so, given the project is being subjected to scrutiny because of its risk and cost profile at the board level). Sometimes boards seem to feel that the regulations they pass that apply below them need not apply to the board itself! If this happens, make sure that you address this deficiency directly, professionally and quickly.

If the board or individual board members are to be directly involved in projects, it may fall to the CEO or CFO to help guide this relationship and its boundaries. The CEO and chair of the board should engage in proactive and early dialogue about what role the board or board members will play, how to make sure this remains well within normal organizational practices and how they will each support the project manager in dealing with any potential conflicts. CEOs especially must work to create a "safe environment" where project managers can discuss their concerns and seek advice on how to proceed, especially when the stakes may be high for all concerned.

THE BOARD AS PROJECT MANAGER

As frightening as the notion of managing a project by committee may be to anyone who has experienced it, there can be times when this role is assigned to the board, or sometimes to a particular member of the board. There are many reasons why this decision may seem appropriate: high-cost or high-profile project, lack of experienced resources or funds necessary to procure project expertise, or the specialized knowledge or role of a board member.

However, in our experience, few of these, either individually or in combination, can justify this potentially risky decision. We actively encourage CEOs we work with to avoid it. The instant you make the board responsible for an operational project you defeat the very notion of separating governance from management. There are no longer any strictures, controls or second chances.

The only exception we have seen in our practice that we feel is justified is when you have a sufficiently complex organization that the board itself has its own processes, procedures and systems. In this case, the board may have projects related to its own operations (as opposed to those of the organization or company for which they are responsible for providing oversight) and they may even have their own staff resources to manage and assign to their work. This also often occurs in highly complex, multi-stakeholder organizations, such as universities or hospitals, where boards have to deal with a significant burden of community involvement, communications, fiduciary oversight and strategy concurrently.

Even so, blurring the lines between management and governance always increases risk. If the board feels this can be addressed by singling out one board member and assigning an operational role, they may also be jeopardizing the credibility and continuing effectiveness of the whole board and may not clearly understand the nuanced differences between governance and management. This is why the focus of our comments in this particular section are directed towards CEOs and/or board chairs, since these are the people in the best position to avoid this decision—or control the decision once it has been made.

Based on our collective experience governing hundreds of very large and complex projects in both the public and private sectors, the greatest areas of risk in allowing a board member to take on the role of project manager are these:

- A fundamental shift in their alignment with the board at a strategic level
- The ability of the board to hold their colleagues' feet to the fire in the same way they would for management
- The challenge for the board member/project manager to step back into the board role, which may feel more passive now
- A board member will have pierced the veil of management, so to speak, and no matter how well run your operation is, there are disadvantages to being scrutinized at a day-to-day level by a member of your board. While well-meaning, they may identify many other functions in the organization that they feel could benefit from their direct input.
- If the project should fail or a decision is made to cancel the project, the board member/project manager may have lost status and collegiality with board peers and senior management. You could

potentially lose a very good board member, and you already know how hard they are to find.

- Potential conflict of interest. There are clear lines between management/operations and board/governance for good reason. Once you start crossing those lines, no matter how good the reason may seem, you jeopardize the credibility of your board. Imagine the media heyday if you assign a high-cost, high-profile project to a board member, and there are serious cost overruns or the project is canceled after significant resources have been invested. Try explaining that to your funding agencies and donors.

As you can see, we have a very strong bias against the board as project manager and believe that most CEOs and board chairs should avoid this situation. In the event that you are under some pressure to consider this option, there are a few ways to either avoid or manage the decision more effectively.

- Suggest that the board member who is being asked, or is asking, to take on the role of project manager consider being a mentor to a member of staff who will take on the role of a junior project manager. Explain that this will allow them to make a bigger investment in the long-term sustainability of this skill within the organization and contribute to the development of an up-and-coming staff member. The challenge here is two-fold: to select a staff member who can work capably with a board member, and to establish the rules and roles of the mentorship at the outset, with careful facilitation by the CEO at the outset and ongoing check-in by the CEO throughout the life of the project.
- If there is just no way to avoid this decision, assign a committee to whom the board member/project manager will report. This creates some separation from a direct reporting relationship to board colleagues. The committee may have a few key stakeholders as well as two to three board members (one as committee chair). The board member/project manager cannot also act as committee chair.
- The board member may wish to consider resigning or suspending their involvement with the board, at least for the duration of the project. This not only addresses the *perception* of conflict, it effectively mitigates it altogether. While this may feel like a fairly drastic response, we've actually seen this work very well where, after careful consideration and discussion of options, the board member decided to leave the board and head up a critical

project. This person can then rejoin the board, and participate in governance, when the project is completed and no longer poses a potential conflict of interest.

Voices of Experience: A Board Chair

For this chapter we looked to one of the best board chairs we know and asked him some critical questions about the governance versus management responsibility of the board in relation to projects. Denis St-Amour is the former President and CEO of Drake Beam Morin in Canada. He was also CEO of their North American Coaching operation out of New York, and is currently President and CEO of the Cyberna Group in Montreal. Denis not only brings extensive experience in the private, for-profit sector, he has also been the Chair of the Canadian Board and is currently International Board Chair of World Vision, providing governance during times of significant growth and restructuring.

What do you see as the board's role in selecting and monitoring strategic projects?

The board needs to continually and consistently be asking senior management about where the organization is in relation to key deliverables. Working together, the board and management set a reasonable timeline to achieve agreed-upon milestones. In this way the board knows what they can reasonably expect from this project and when, and be able to decide if progress against the milestones, costs and timeline is sufficient to continue. They can determine if management needs to be encouraged to stretch a little more, or if they are in danger of stretching themselves too far. The board needs to be able to measure the work not only in terms of time and cost, but also the substance of strategic outcomes. Management must provide sufficient information to give the board the confidence required to stay on their side of the governance line.

The goals and objectives of key strategic projects, those that are potentially critical to the ongoing viability or growth of the organization, must be clearly reflected in the board's evaluation of the CEO's performance. Clearly defined project outcomes, timelines, milestones, and costs define the project activities that create the ultimate project results. These provide the qualitative and quantifiable measures against which performance can be accurately measured.

How do you determine how "hands-on" a board needs to be with high-profile, high-cost projects?

This requires a balance of both hard and soft indicators. Soft indicators being the credibility of internal/external resources assigned to the project: do the people in front of the project have a "high stock value" with the board? Boards must have confidence in the people they entrust with a strategic project. Understanding that projects are organic and may experience changes along the way should not shake that confidence as long as there is a pattern of high accountability and trust with the project leaders.

It is equally important for the person or team managing the strategic project to know that they have a level of support from the board that, should they find themselves navigating stormy waters along the way, they can focus on steering the ship and not be distracted by the need to look over their shoulders.

Is there pressure on boards today to cross the line into management functions because of increased liability and media attention? How do you draw the line?

Clearly there is. You need only look at some of the high-profile trials resulting from governance failures featured in the popular media. Very successful and well regarded board members came across as only doing superficial diligence, leaving themselves open to significant criticism and potential lawsuits. More and more, board members are expecting that if their names are attached to organizational governance, then they want access to the level of information required to support good decision making. It is also critical that the board have faith in the CEO as someone who is fully transparent and accountable and not just someone who wants to use the board as a rubber stamp. This can create tensions if not properly managed. Today, more than ever, boards and CEOs must understand the need for "a spirit of ownership and partnership" and that everyone involved has something at stake, something for which they are accountable. "Well run organizations have Executive Limitations set in their Standing Policies for both CEOs and boards." (As one person once said to me, "Just let me know what the rules are and I will follow them.")

Many organizations have great governance models on their books, but they don't practice them. It is the responsibility of the board chair, with support from the board development and/or board nominating committees, to design and evaluate governance standards for the board. It is up to each individual board member to be knowledgeable and apply the standards to every decision they make, and for the CEO to embrace the role of board governance as not only necessary but value-adding. Ignorance is no longer a reason; it is only a poor excuse.

What are the challenges to boards governing in a more project-based environment?

The board needs to stay focused on end results. What is the committed end result and how does it link to strategy are key questions to ask. Often projects report into a board committee or task force and this can be challenging for everyone to manage. The board needs to set expectations with the committee on how they will measure the progress of the project against goals and objectives. Because the committee is the first line of project reporting, it is important that it is able to identify issues that could derail the project early enough to take action or escalate as required.

The board must also deeply know their standing policies and practice them faithfully. Key projects should be subject to the same level of scrutiny as any part of the operations of the organization for which the board is ultimately responsible. These are the play book to be used in every meeting in relation to the various committees and portfolios to guide board activities and decisions and help them to determine if what they are being asked to do is governance or management—and to guide them in sorting out any conflicts to ensure they remain on the right side of the line. In every meeting, the lens to be used by the board for every activity of decision should be: is this Governance, Policy, or Management?

For more real world stories, check out www.projectgurus.org.

FINAL THOUGHTS

When working with boards in relation to project management, the stakes are high. The Triple Constraint Model (Scope, Time and Cost) or even a Quadruple Constraint Model (Scope, Time, Quality and Cost) can be insufficient to guide the project manager to meet the board's demands around clarity of expectations and access to information. In fact, we have been unable to even locate much in the way of a model for project managers to follow in terms of interacting with the board.

This may be, once again, a reflection of how infrequently the profession tends to think of itself in strategic interactions. In our coaching and staffing practice, we often encounter project managers who have never, or infrequently, presented to their organization's senior management team, let alone the board. However, in the not-for-profit sector, we find the opposite is true, and those working in this context are frequently in contact with the board. So while we do not believe that generalities always work, we propose a six-dimensional model for project managers to follow so as to create confidence in their interactions with a board of directors. It looks like this:

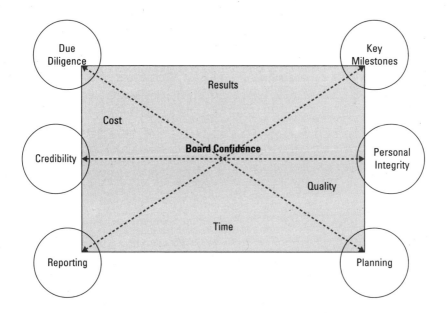

A Six Dimensions Model of Board Governance for Projects

Based very much on insights gained from real board members (as featured in our "voice of experience" panels), this speaks to the need for project managers to ensure that they have done an appropriate amount of due diligence on what is being presented to the board before it is presented. Make sure you anticipate board questions and see their likely concerns from a governance, rather than management, perspective. This creates credibility, an essential ingredient supported by a demonstrated sense of personal integrity. Finally, reflect on exactly what kinds of reporting the board wants. Normally this involves the basics of briefing them on the plan and its key milestones so that they can monitor strategic risk and reward at the enterprise level.

With the board as your primary audience for project approval, reporting and evaluation, you need to broaden your project management understanding to include a solid understanding of modern governance and your organizational context. You can use our Six Dimensions Model of Board Governance of Projects as a starting point for that exercise. It provides a framework so you can locate the information they need to achieve their governance requirements. The model also identifies the most often-cited "soft skill" areas that boards want to see in order to gain the level of confidence required to keep the board on the governance side of the equation.

As project management becomes more and more strategic to the enterprise, we assume this will lead to more and more interactions between project management professionals and boards. From this, we anticipate that this model is only a starting point for the potential to define how our profession fits into governance models that will surely emerge in the future.

## CHAPTER SEVEN

Good Leadership
in Foggy Conditions

Co-written with Carol-Ann Hamilton

Carol-Ann Hamilton is a noted speaker and co-author of the best-selling book The A to Z Guide to Soul-Inspiring Leadership, *written with Dr. James Norrie. She is in private practice in Mississauga, Ontario, as a coach and consultant. As an expert on new modalities of leadership, in this chapter she brings her wisdom to the topic of how leadership style and change management techniques can make a difference for project managers and executives trying to move their organizations forward in adapting a more strategic perspective.*

Getting any kind of work done in an organization requires two fundamental skill sets: the oft-lauded "hard skills" and an accompanying array of "soft skills." Since this book challenges leaders to take initiative around strategic project selection and portfolio management, making this endeavor work requires that you play at the top of your leadership game. That's what this chapter is all about. When you are asking people to change the way they work, leadership style matters. Unless you can successfully engage others around you in the need for change, you risk failure.

As co-authors, James Norrie and I have worked together for over a decade. In our experience coaching hundreds; of executives and senior practitioners around the world, we believe that technical and problem-solving (or hard) skills can be taught and acquired; but it's far less intuitive to acquire attitudes that underpin interpersonal relationship (or soft) skills. So often, even in an emergent field like project management, the balance of leadership preached by the experts seems tipped to the superiority of the visible and tangible signs of analytical leadership over the less-understood soft skills. This common error leads to failure when we focus on the "what" rather than the "how"! So in this chapter we will dissect the necessary leadership traits and methods required to make PPM (or any other complex strategic business process) work from both perspectives.

In the long run, "results at any cost" projects deplete performance. This is also the case with an overly aggressive portfolio of projects that are too big, too loose or too ambitious. That is not to say leaders cannot and should not *challenge* their organizations to perform. They should. Rather, while any project in the portfolio can potentially achieve superlative results, if it's at the hands of a beleaguered team, shoddy practices and questionable ethics, how fulfilling (let alone profitable) can it be at day's end? Longer term, this approach to project leadership is rarely sustainable.

Borrowing a page from the realms of organizational performance measurement and management, just as it is damaging in the long term to focus exclusively on financially-based metrics, so too is it naïve and dangerous to assume that "*what*" is achieved will eventually override "*how*" results are attained. Since the intent of a balanced portfolio is sustainable competitive advantage, if this is negated in the short term by ineffective leadership practices, the result won't be optimal. This is something for project managers especially to be concerned about—the profession often seems to say that it is only the output or results of what you do that matters. We disagree.

Concentrating on the "*what*" and the "*how*" together helps great leaders realize balanced organizational performance. In fact, if there is a bias in terms of what to pay more attention to, we think the "how" will always prevail. Why? Because if past project experiences have left a bitter aftertaste among your team (perhaps through a poorly executed "how"), it's a sure bet your talent will run as fast as possible in the opposite direction should a similar opportunity to participate arise in the future. Talent is precious to any organization and it will not stand for this for very long. In today's globally competitive environment for employees, no organization can afford to support leaders who deliver the results but at the expense of their people.

No matter what role you play within your organization (board member, executive, project manager or contractor/consultant), you have picked up this book because you want to make a difference. So let's begin our critical thinking around leadership during times of change with a look at your personal leadership style.

PERSONAL LEADERSHIP STYLE

Have you ever noticed that some people can hold a lofty title without acquiring followers? On the other hand, are you aware of examples where someone holds a nondescript title or middle-level position while inspiring everyone around them? Leadership opportunity is not position dependent. We always tell those we coach, "Anyone can lead from anywhere in the organization."

What creates this magic? The answer lies in your personal integrity, attitude, vision and values. It's through your personal style that others accord you credibility and trust. Positional power simply doesn't cut it; personal power does. That's why we possess little patience for executives and project management professionals when it comes to their typical laments: "I get no respect," or "Why don't they get it?" or "I can't influence outcomes because I lack authority." These lame excuses indicate an absence of personal leadership.

So we challenge you to examine your inner world first and consider *who you are* when you lead. What do you project when you are front and center? These are crucial moments when you can demonstrate impact as a leader, but the real moment of truth is actually found when nobody is looking. In these moments what do you notice? What kind of trail do you leave in the wake of your daily activities—projects and otherwise? How introspective are you about not only what you accomplish but how you accomplish it? We challenge you to reflect if people and your relationship to them really do matter to you or if it really is more about the results they deliver.

If you spend even a minimal amount of time on this kind of reflection, we're certain you will soon find that personal leadership lies at the heart of ultimate project success. In fact, this notion of personal leadership—first and foremost—lies at the heart of all success in life.

WHY SHOULD YOU CARE SO MUCH?

Emphasizing the importance of personal leadership in terms of how performance is achieved may sound like corporate idealism; perhaps you are asking yourself, "So what?" and, "Why should I care?"

Rightly so . . . except for one thing. Demographics will soon catch up with organizations everywhere and result in a potentially massive exodus of talented employees that will overwhelm growth and productivity, unless you become and remain an employer of choice. When coupled with radically shifting expectations among the next generation workforce, these trends will force all industries to face a staffing crisis of a magnitude never before seen. To see how this might look in practice for you, consider any part of the world that has an "oil patch." The mad rush to extract oil from the ground drives huge salaries, even in a restaurant or coffee shop; gives rise to recruiting drives to bring in more of the most basic labor skill (while offering free housing and relocation bonuses); and creates a local market where retention of every single employee matters, because the costs and time it takes to replace them affect production.

True, you've heard that before. But make no mistake about it, while it was once the case that you likely had more people looking for a position with your organization than you had positions to offer, do you now notice fewer applicants than in the past? Or perhaps you get lots of applicants, but few who are qualified? Soon there will not be 10 employees in line to replace disgruntled departed team members.

Or maybe you have noticed the never-ending abundance of "Help Wanted" signs in the retail, food service and hospitality industries these days? The deluded belief that "we can always get more where they came from" has become a complete fallacy. More than ever, your organizational leadership practices have a meaningful impact on results through the retention and performance levels of employees.

Key employees will vote with their feet if they do not believe their work matters. Many are leaving legacy corporations for start-up initiatives, where their cry for character-based personal leadership and meaningful work can be answered. This plays right into the PPM process you have already learned about, for it alone can help establish that the work being done is strategic and can make a difference if completed. This kind of process also speaks to the ability of *anyone* in the organization to lead by creating and submitting winning project proposals that do not depend on political sponsorship, but rather on a demonstrated connection to results to get approved.

In fact, properly implemented, PPM unlocks the ability of your talented employees to dream up incredible solutions to problems that your competition can only admire after the fact. Since there is lots of research to support that meaningful work is even more important to employees than money or security, projects more closely tied to strategy will make the work of project staff more meaningful and enhance their feelings of making a difference, while also making it easier for executives and project managers to be inspiring leaders. This is truly a win-win outcome.

In this new workplace, having the means of execution makes all the difference. This suggests an emphasis on a satisfied workforce. This is the "secret sauce" of an agile organization, able to prosper in a competitive global environment—versus an organization where miserable morale creates apathetic employees whose low productivity can be seen in skyrocketing project costs, missed deliverables and strict adherence to minimum expected performance levels. In that organization, it would be impossible to execute the optimal portfolio, even if it were possible to establish it in theory. Motivated and engaged employees will not only help you pick the right portfolio, they will be eager to execute it as well!

There you have it—the business case for fostering the *how* of project performance through your own inspiring leadership—no matter what your role and responsibilities. So now let's consider the specifics of what good leaders do to create this progressive climate within their organizations.

A NEW LEADERSHIP MODEL

Our direct experience with CEOs confirms they want project managers' valuable store of skills and expertise, especially because of the perceived value of what the discipline can contribute to the bottom line. Yet executives often find themselves confounded by these same professionals' inability to articulate the essential value of portfolio and project management and its direct role in strategy execution. There is a lingering sense among CEOs that perhaps project managers don't really understand the business and the choices it must make.

At the same time, project managers are intensely frustrated over not being fully engaged in the strategy making and execution processes within their organizations (as we saw in the early survey results in Chapter 1). Oh, the railing we hear about senior executives who do not support them—it can be heard far and wide in any meeting of project managers anywhere in the world. Horror stories abound about bad executive sponsors, unsupportive CEOs and the inability of the organization to grasp just how difficult project management jobs really are. Have you ever found yourself saying anything similar? This means both sides of this important endeavor are frustrated.

So, what can we do to reconnect these two apparently disparate camps? Is there a leadership tactic that can help both executives and project managers communicate better? Yes. A desire to *get it right* more than a desire to *be right*! For executives, it's about strategy execution—getting the right people doing the right things to remain competitive. This requires a willingness to demonstrate new leadership paradigms that encourage risk and reward concentrated effort. For project management professionals, it's about *doing it right*, without falling into the trap of focusing their organizations on the wrong things. This suggests a natural alignment between the goals of the CEO and the strategic project manager's desire to partner to support these goals. It further reinforces the relevance of PPM as a method of creating this partnership and sparking necessary dialogue between the two groups. This should make each party (executives and strategic project managers) receptive and comfortable in reaching out to the other for help.

In our collective work on hundreds of projects executed globally, we have found a common theme among organizations that seem able to perform well and accomplish work fast. Over time we have researched the common aspects of leadership we find in these organizations and have distilled them into some simple leadership lessons. Nine of these appeared as "Simple Leadership Equations" in our first book.

However, we have discovered even more about this important topic since that first book was written. For instance, the call to leadership action for executives requires a willingness to set aside some of what you know (or think you know!) and instead focus on inviting project management professionals and other employees to come forth and present their expertise in novel ways that support your business goals. This often requires you exhibit a new set of behaviors that is more about listening than presenting or speaking. In our coaching practice, we help often aggressive and action-oriented executives focus on three specific "soft" competencies we call the triple A's: *ask, acknowledge, and activate*.

When we coach project managers and other aspiring leaders, we focus on their willingness to risk courageous conversations with senior leaders and a readiness to set aside any sense that they are not full partners in the strategic management process. For them, we focus on three slightly different "soft" competencies that we call the triple C's: *courage, confidence and change*.

But what is the "B," if we have only covered the A's and C's? The cloud in the middle represents changes in *behaviors* that are the outcome of both sides of any leadership transaction applying the A's and C's together. By adopting an attitude of mutual discovery and respect, we challenge both sets of professionals to consider how their perspective and behavior may be impacting their results. This framework often helps our clients evolve towards new ways of interacting and working together.

The ABCs of Leadership

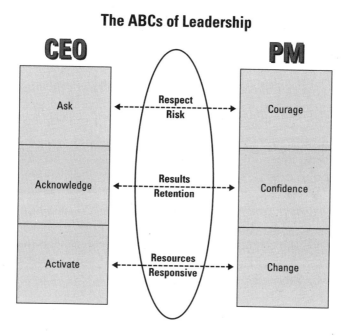

Let's consider specific examples to see how this works. If the CEO expects employees to be honest about the challenges that are preventing them from optimally executing strategic projects, he or she must be willing to ask for this input directly. By asking for and inviting courageous conversations, employees may be more prepared to take risks and engage in an open, honest dialogue with their leaders. The result is you hear what you *need to* hear rather than what employees think you *want* to hear. While we often mouth the words that this is what we seek as leaders, when we negatively respond to employees when they take the risk of actually telling us what we need to hear may not create a climate of trust and respect. So they revert to telling us what we want to hear, and feel better about the response they get—but meanwhile the organization is operating at sub-optimal levels. Executives will eventually pay the price, in diminished results, for this deception.

If we want employees to take bold risks, sometimes they will fail. Our response needs to be as consistently supportive when we collectively fail as when we succeed. A culture that engages employees fully and invites them to take risks to generate results is one that can retain and attract talent, a key component of competitiveness. Here again, we see the direct connection between soft skills and creating a winning culture and climate.

Finally, as we chart the course for change and activate projects that will help shape our destiny, we need employees and project managers who are willing to change. This is not always easy given the natural human preference for the known over the unknown. Good leaders provide the resources required to engage employees in the change process, resulting in a culture that is responsive to and ready for change. This is a must-have for most CEOs and other C-level executives if they intend to thrive in today's competitive world.

As with so much in life, the more you put in, the more you get out. There is no return without risk. This applies as much to senior leaders as it does to project managers or other employees. For many of you, engaging in this type of open dialogue may be new and initially seem risky. But in our experience, the outcome is absolutely worthwhile for everyone. So let's spend some time delving more deeply into the specifics of these critical behaviors and learn more about how they apply in action within organizations every day, every way.

THE THREE C'S FOR PROJECT MANAGEMENT PROFESSIONALS

Courage

Courage is about maintaining your convictions even when the going gets tough. Courage is not the absence of fear but proceeding in spite of your fears, and courage is not a choice—convenient when you have no "skin in the game" but abandoned when things become contentious. Courage is about standing up, and no matter what the circumstances, and telling the truth. People often embrace this trait in personal lives but seem to resist it in an organizational context, perhaps because it's easy to link courage with being confrontational. We do not make that association. Professionals and executives can learn to be courageous without being overly assertive or aggressive.

In the context of project management specifically, courage involves deciding not to hold back what needs to be said—being candid, even if it feels perilous. When difficult messages have to be delivered, courageous project managers are prepared to go eyeball to eyeball with the most senior leaders, taking the heat of controversy. They value honesty—always.

To illustrate aspects of the three C's, let's consider the story of a professional services firm where a project involved truly "bad" client behavior. This had been permitted to go on for months before someone finally chose to do something about the situation. Now, everyone knew what was going on. But the only reason everyone colluded in not calling out what we call "the elephant in the middle of the living room" came down to a single factor—the huge billings that were at risk. Everyone chose to turn a blind eye for fear of jeopardizing the supposedly sacred contract. They felt a need to preserve revenue at all costs! Yet, *someone* had to do something or the project was ultimately going to fail.

Never mind that the client regularly missed deadlines, while the organization delivering service was expected to respond, "How high?" on the command, "Jump!" Never mind that the client insisted its advisors were always "wrong" and made them all work overtime at their every whim, wreaking havoc with work–life balance. Never mind that a key player from the client team kept attempting to "triangle" one of the service team members into her private agenda to make the project manager appear downright stupid at every turn, especially when he showed the remotest signs of exposing her shenanigans.

With only this much information, you can imagine what a challenging dynamic this set up for the new project manager, who was appointed with the onerous task of cleaning up a really messy project. All we can say is thank goodness for the considerable seasoning Marianne possessed in dealing with this turnaround situation. Without a doubt, she needed to draw on every technique from the depths of her experience to engage in the tough conversations that desperately needed to take place with her client. She was not willing to become part of "everybody"; she wanted to be the "somebody" who finally made a difference. She had the courage of her convictions, based on years of experience and her intuition that something had to be said and done to fix the situation.

Courage in this instance meant refusing to set aside what needed to be said. Her commitment was the polar opposite to the typical avoidance behaviors we often see in organizations. Having performed her due diligence by obtaining the input of all parties involved, she knew it was time to stop the game and get on with the tough job of delivering a key project on time and on budget.

Marianne was courageous without being arrogant, insolent or abrasive. As her first step, she took a gentle approach of conveying necessary client performance improvements as "areas of opportunity" that could be "worked on together" to produce "more strategic results for both of us." This was very savvy positioning that conveyed to the client empathy for the situation while still demanding action to improve gaps. She framed her insights about the existing situation as a need for "continuous learning" or "improvement." By adopting a more neutral and collegial tone and tenor, she avoided the defensiveness that so often results from the more typical "let me give you some constructive criticism" method that some leaders rely on too much. Whether critical or not, criticism is rarely a welcome event in most people's view!

How did it work? Well, never as well as Marianne hoped for, and it didn't fix all of the problems the project faced. But it was a good start. Now, this is the point where many project managers we know might have stopped for fear that any additional action might spark fears of reprisals for "going too far." Not Marianne. There was more work to do to get this project on track.

Faced with the client's continued refusal to acknowledge subtly damaging behavior that was still negatively impacting critical project outcomes, our project leader shifted gears. Relying on her experience with the ABC leadership model, she approached the CEO about the problems with the project and the behavior being demonstrated by key leaders he had assigned to support it. Some of you may be recoiling in the misplaced view that she was acting inappropriately; however, remember our definition about courage? The only path Marianne felt she could take was to the top. So that's where she went. The conversation challenged the CEO to get engaged with the project directly; to see for himself what was going on and to take responsibility for the outcomes of something he had promised investors would make a difference to the company's P&L. This artful re-engagement of the CEO did ultimately uncover bad executive behavior that he was able to correct and, over time, the project began to recover, to everyone's relief.

This gutsy project manager stopped at nothing to bring in a winning project outcome, despite the potentially grave consequences of having the client disband a million-dollar-plus initiative with her firm. Marianne was prepared to go to the wall on this one because it mattered to her personal integrity, but also because of her belief that if the project failed, the impact on her firm would be as negative as if the project were canceled immediately. So why not risk trying to get it back on track now?

In the end, her firm's leadership backed her up completely. Their combined efforts eventually led to the changes in the project leadership on the client side that everyone could see were required. But it took somebody to make it happen! Marianne recognized that a successful relationship is not always about being nice but is founded on being real. As a result of her courage, she has developed a strong reputation for being trustworthy and a truth-teller. She has gone on to lead several more project turnarounds, always with the same view of leadership—doing what is right for the project and thus creating winning outcomes, regardless of what is best for her personally. This is not easy to do, but it's the right thing to do.

Confidence

Have you ever noticed how many project managers and others often look outside themselves for solutions—as if others must hold better or smarter answers than they do? Frequently in our practice, project management professionals will ask what we think they should do to solve a given problem. It would be one thing if they led the conversation with what they planned to do and sought some feedback from us, but they don't. They want us to define the solution for them. We find this lack of inner confidence fascinating.

Our tendency in these instances is to turn the question right back to our clients. We use a single but powerful four-word sentence: "What do you think?" If they still answer, "I don't know," we reply with, "But if you did know, then what would you think?" When we invite project managers to consider this question more deeply from the perspective of what they *do* know, we inevitably find they come up with at least the start of a great answer. Indeed, if any of us searches long enough, we will actually discover we often have the resources within to deal with almost any challenge.

Here's another secret, one that some consultants might not want you to know: the answers our clients generate are frequently superior to those we would have provided. One of the things we pride ourselves on as a firm is helping clients find their own solutions. What we bring to the table is process, techniques and tools that are proven to be useful in the pursuit of optimal performance. But the definition of how to perform is still best left to our clients, who simply use our tools to help themselves get there quicker. This approach brings the added benefit that the teams we work with buy into their own solutions—because they generated them!

Humans are powerful creators of their own reality, and if we believe in ourselves, we can do great things. If, on the other hand, we focus on what we can't do or don't know, the impact is exponential erosion of our confidence. This prevents us from achieving our full potential. As Henry Ford is said to have once remarked, "Whether you believe you can or can't do it, you're right."

We can easily apply this directly to the project management world. Today's aggressive business climate is not for the meek and mild. Just think of what would have happened to the project in our "bad behavior" client example if Marianne had come in all apologetic and self-effacing. She would have been crushed!

Clearly, solid project leadership requires strong elements of personal confidence and this is derived from a strong ego. The counter-balance to ego so that it doesn't rage out of control is personal insight. When ego drives project managers to exude confidence, it is working in its best sense. The perceived confidence a leader demonstrates makes for willing followers. It helps those around you to focus on results, because they sense you know what you're doing and their efforts will make a difference.

However, too much ego or self-confidence leads to arrogance. This reduces your potential to truly *listen* to others and *hear* the truth. If this happens, you have lost sight of the need to believe that others may also have the right answers. While you may aspire to do what's right, this does not always make you right. This is a critically important nuance for any project leader to understand.

And the feedback we've received from countless executives during this five-year study echoes this observation. Senior executives do not need more "yes men." They already feel isolated enough at the top. What they really want is to be told the truth—respectfully and insightfully. You are where you are because of your specialized skills and knowledge, and you have important strategic contributions to make. But you have to be willing to confidently make them.

More than any other profession we should be able to understand this, because our work involves "unique endeavors with a defined start and finish." In most organizations, the business environment of yesterday is solidly in the past. The future is not yet completely clear but it is the most important organizational project we have to work on. Strategic processes like PPM can help.

As our work environments get filled with more and more information (but perhaps less wisdom!), what we know to be true becomes critical to slicing through the fog. Gone are the days of waiting until all the information was available before making a decision. Business at the speed of thought is becoming more the norm. The result is that we are forced to rely on our own centers of knowledge and know-how to navigate in periods of uncertainty. Act as if you can't fail and confidence will follow. And that's what we think good project managers are really good at!

Change

This last concept could just as easily be called imagination and innovation, because it involves acting as much on intuition as it does on information.

During our years as consultants, we have discovered several consistent differences between ineffective and effective project management professionals. The best project leaders are acutely aware of the need for innovation that is founded on a sense of imagination about where to go and what to do, even if it's never been done before. This skill is hard to define and as a result it doesn't exactly make it into the *Project Management Book of Knowledge* (PMBOK), which many project management professionals rely on to provide them with guidance in their field. However, the best project managers we have worked with embrace change as a challenge, while run-of-the-mill project managers consistently perpetuate the status quo.

We see this with PPM implementations. While we know that every single organization with even a small number of projects runs into issues of prioritization and a continual struggle to stretch their limited resources over too many project requests, we are still startled by the number of professionals in our field who do not believe we have any role to play in strategic decisions about how to resolve this. They wait patiently to be told what to do from the sidelines, believing this issue is one for the executive team to resolve. What nonsense!

In our opinion, informed by years of working with a number of outstanding clients and their project teams, it's actually not appropriate to claim full effectiveness as a professional if all you do is "manage projects." Even by their very nature as the building blocks of strategy, your role in making these projects succeed is strategic. Furthermore, you alone have the informed knowledge and skill to help create and define *workable* projects that have the potential to succeed and make strategy real. You are one of the most vital lynchpins of strategy execution within your organization, even though you may not have realized this.

The pride that traditional project managers hold in their ability to think logically and rationally (as opposed to innovatively) is precisely what may ultimately prevent them from becoming strategic. Rigid insistence on "doing it right" is a surefire recipe for igniting the ire of any seasoned executive whose focus, by definition, must be on "doing the right things."

By imposing unrealistic constraints on the way we would *like* organizations to operate (for our structured convenience) as opposed to the way they actually operate (in a messy, unpredictable and chaotic way), project managers as a whole may be bringing upon themselves that which they dread most: being reduced to a pair of hands that is simply told what to do. Executives must roll their eyes sometimes when we enter the room. An over-emphasis on structure and order above all else lies at the root of unworkable solutions and confirms their suspicion that we are ignorant about how business really works! Yet this trap is largely avoidable in the first place.

Instead of expending such considerable energy insisting on the logic of getting projects done right, how about lifting your sights to the organization's broader vision and making sure we're doing the right things? By offering your expertise in ways that spark the executive imagination and create unique solutions to real business problems, messy as this may seem and as complicated as it will be to do, the perception of your value and the value of our profession increases enormously. For example, one of our clients included a project management staff member in IT who had a suggestion for a simple approach to automate certain aspects of initiating project requests. However, the CIO was seemingly focused on "enterprise-wide project management systems" and so the PM worked diligently on finding the big, expensive but comprehensive solutions the CIO seemed to want. After all the time spent researching the big, bold solutions, the approach that was ultimately authorized was from a home-grown set of automated tools in Excel and Access.

It took longer to get there because the CIO didn't ask for creative solutions, and the PM was not courageous enough to gently suggest alternatives. So much time wasted with each party thinking they were serving the other, but who was doing what was best for the organization?

Change-oriented project leaders invite those around them to imagine the possibilities (and not always the probabilities) of what would be different if a project succeeds. And more and more we see the literature in project management dealing with this issue of connecting project vision to organizational mission through tangible measurement of outcomes. Furthermore, truly future-oriented project leaders lead the charge in redefining the status quo. They derive tremendous empowerment from transforming "how things have always been around here" towards "the way we *want* things to be around here." They are go-getters who make good things happen. They create climates where people want to participate fully.

By mentally moving themselves from a reliance on predictability and overly rigid methods as sources of comfort to an attitude of self-control over change, leading-edge project thinkers also help reduce anxiety about change and its impact on everyone around them. By not perceiving change as scary, they help others to also see change as an opportunity.

THE THREE A'S FOR EXECUTIVES

Ask

Contrary to our advice to project leaders to speak their truth with confidence and courage, we implore executives to know as little as possible. In fact, the fewer answers you have, the better! For some of you, this will immediately resonate and confirm what you have always known: speak last, but speak well. For others, especially if you have a less-developed sense of your own leadership style, you may feel that appearing in control and having answers is your destiny and what others expect of you—especially as the CEO.

We realize most executives are schooled on the job. Sure, you may have an education, perhaps some certifications or affiliations. But what you really rely on to get and keep the top job is your experience. Called upon to make instantaneous decisions about a multitude of things every day, you may eventually develop the mindset that you should know everything.

In *The A to Z Guide to Soul-Inspiring Leadership*, we labeled this phenomenon the "myth of the mighty." And the myth is not one-sided: followers want to believe their leaders have all the answers. But realistically, they cannot, and we should not expect them to.

Given the hyper-speed of business today, the best leaders understand that the only way to get often elusive answers is to welcome input from a wide variety of sources. In fact, the Information Age has given rise to the instant availability of so much information that the challenge of today is to cut through this foggy clutter and turn it into leadership wisdom. Where in the past there were perhaps only one or two sources of accurate information to consider in a decision, today there may be hundreds. Since we cannot slow down to absorb them all, compare them and then determine a "best-fit" decision, the real skill of leadership is to narrow down the sources of information, focus them rapidly on a suitable course of action and, as we say to our clients, execute the decision with precision. It's the only way to survive today.

This is why we advise project management professionals to be inwardly more confident, so as to proactively offer up their own answers! Can you see how it all fits together? Just as you are desperate for others in the organization to help you determine how to proceed, they desperately want to contribute to cutting through the fog and getting on with it. You can become incredible partners in the execution of strategy if you work together. So ask them for help and make sure that they can confidently and courageously tell you what you need to know before they make the decision, not after the fact.

Smart executives set aside personal biases to focus on getting good information quickly. They often find that this information is just below the surface of the organization and seems ever so slightly out of reach. We know this from countless hundreds of hours coaching CEOs on improving organizational performance. The real answers, both in terms of what's actually going on today and what could go on tomorrow in your organization, lie with frontline employees. Yet your most ready source of information is filtered by layers of management, who may be inclined to tell you what you want—rather than need—to hear.

The Three Levels of Listening

The best CEOs place as much emphasis on what is *not* being said to them as on what *is* being said. Listening is the counterpart to asking. Yet, when most of us are interacting, we listen to what is going on inside our own heads rather than observing the reactions and behavior of those around us.

Level 1 Listening

Many of us even self-centeredly prepare our responses before the other person has finished talking! Effective listeners actively suspend this habit and really pay attention so they can hear what the other person is saying. They focus on understanding both the content and context, trying to locate the perspective of the speaker. Still, they miss subtle cues and signals. While they are focusing on the words as they are spoken, they may miss verbal clues and body language that are also critical to a complete dialogue.

Level 2 Listening

So the next achievement is to not lose the concentration of level 1 but to add a deliberate awareness of the subtle body language and verbal clues that help expand your understanding of how the speaker is feeling while speaking the content. This adds empathy into the context and may aid your understanding of the content.

Level 3 Listening

At this third level, we integrate both the context and content into our listening consistently. We learn to attend to the emotional and behavioral climate around the words we hear. We target this skill for CEOs who participate in our coaching programs. While the project team says, "We can deliver on time and on budget," what they really mean is something along the lines of, "We are really scared because we haven't done this before but will do our best with what you have given us to get it done as planned."

You hear the words "We're confident," but perhaps the body language, verbal cues or other signs do not fit with a confident delivery. Or perhaps there are not confident answers to questions you might pose. Not questions designed to elicit a "correct answer"; rather, you get at the truth by asking questions that invite people into a safe and comfortable environment where they can feel supported to say anything on their minds. You want to discover the truth about what they are saying and feeling.

For instance, if only the leader of the team is speaking, try to draw out the team members. People who believe in what they are saying are excited, passionate and normally unrestrained in their delivery. If you're the CEO, you need to notice these subtle nuances and determine the degree to which your project leaders and teams really mean what they are saying.

Now, this is not to suggest that anyone is being duplicitous or evasive. In our experience of analyzing over 100 major project failures after the fact, the project teams involved did sincerely believe they could succeed. But we found that common to the projects were a large number of unspoken reasons for why the projects failed. These could have been avoided or at least uncovered through a more open and honest dialogue at the outset, during the project proposal and approval stage.

The critical skill here is between asking questions to obtain data versus queries designed to unearth personal feelings or needs. The latter offers authentic clues about others' state of being or motivation (level 3 listening) while the former yields only analysis, reasons, explanations and justifications (level 1 listening). One leads to interrogation and builds defenses around the "right" (or rote) answers, while the other removes barriers and creates powerful relationships that lead to valuable insights.

Marvin: A Case of Poor People Skills

Here's an example drawn directly from the co-authors' 50-plus collective years of working within corporate Canada. Marvin was a very successful, well-paid project manager who was considered the "go to" guy for tough projects inside his company, a large financial services firm. If there was one thing Marvin was known for as the leader of his organization's project managers, it was his "gotcha" questions. His team notoriously let newbie project managers learn this trait the hard way; there was a feeling that getting through your first "interrogation" by Marvin was a rite of passage into full team membership.

This code had developed over many years of successful project delivery where Marvin had convinced the CEO, and likely many of his own project managers, that his skill in spotting possible project risks was vital to their track record of success. Yet this ability to trap others was actually born of a personal need to feel superior to others because, deep down, he was quite an insecure person.

The organizational consequence for this behavior meant that his team did not take risks and rarely, if ever, involved him in problem solving or creative brainstorming. As it turns out for both Marvin and the CEO, this created over time a kind of "me too" culture within the organization, where projects began to bubble up mostly on the basis that a competitor was "already doing this" or "seemed to be moving in that direction." Never shy about asking for "the facts," Marvin made sure he knew what was going on in his industry and kept the CEO advised of how his projects were tracking compared to the competition. There was a sense that the organization was "keeping up" as a result.

What they were not doing was getting ahead. And in a global world where nothing stands still for very long, over time this made them vulnerable to a takeover. While valiantly fighting it off, the CEO eventually lost board support and the deal was approved; the CEO was let go. In retrospect, he realized he had been too focused on current results and had lost sight of the future. This is easy to do in a publicly traded company, where the emphasis is so often on the latest quarterly forecast and comparisons to last year's numbers. But the best CEOs realize that in spite of all this pressure, they must innovate. The speed of innovation and their ability to bring new products and ideas to market is really the best competitive capability they can create in their organizations.

By not releasing his critical nature, Marvin teaches us what can happen when leaders ask questions from a place of blame, fault-finding and criticism. Mired inside a vicious circle of right and wrong as defined by the leader, project teams narrow their sphere of influence to that which can be aligned with and controlled by their leader's expectations. They become microscopic—focused on *answering right* rather than on discovering the *right answers*.

Marvin achieved his individual goals and targets over many years and was paid significant performance bonuses by his company. In fact, if we took a peek at his performance appraisals, we would find a leader who excelled in the "what" and was seen by the CEO as an important contributor to the organization's achievements. And Marvin was—if the achievement was continuously promoting underperformance and being acquired by a competitor.

There were other warning signs the CEO missed, and they were brought to light only retrospectively. Very few of Marvin's team stayed for long. That is not to say they resigned from the organization, although over time a few did that, too. What we discovered while researching this case was that he had an incredibly high rate of internal turnover. Employees would apply for and be hired into project manager jobs, but they had an average tenure of only 22 months. And over the course of five years, 100% of Marvin's team had left.

If these figures don't startle you, it may be because you're thinking a little bit like the organization did at the time. First, some degree of turnover is inevitable, so the few departures of people for other organizations did not provoke any alarms. However, a good exit interview process (which this client has since instituted) would have indicated a pattern of poor leadership and a sense of desperation from seeing that behaviors which really disturbed departing employees were actually being rewarded by the company.

The other reason no alarms were raised is that the turnover was internal. In fact, Marvin was overhead one day telling a colleague about how proud he was of "developing" his team and the "great jobs" they were going to elsewhere in the company. And it was true: his team was seeking out new career opportunities at a massive rate, mostly as a way of escaping him! In our research for our last book, professionals who reported working for a "toxic boss" were up to four times more likely to either be actively looking for, or passively entertaining, new job opportunities. Whether they move internally or externally, unless the move fits into an overall career plan that makes sense, it may be motivated by bad leaders.

With the coming labor shortage, no corporation can honestly afford leaders who will not ask and listen well. So in order to attract, retain and

engage high-potential talent from the next-generation workforce, dysfunc-
tional leaders like Marvin need to be uncovered, recovered or dismissed.
As the CEO, you have a vested interest in making sure you are asking the
right questions and getting the right answers before it's too late.

Acknowledge

As we have seen, it takes more leadership courage to practice the soft skills
(*how* work is done) than to rely on the more familiar hard skills (the *what*
of business results). But it doesn't have to be difficult.

Sure, you can push hard to impose your views on others, rather than
step back to both hear and understand their perspectives. Sure, you can
choose to listen to your self-importance rather than demonstrate powerful listening skills (including silence) and asking sincere questions. Your
power will carry you to the finish line on a short sprint, but will it help
you win a marathon?

Our second "A," *acknowledge*, is not all that challenging, but it does
take time. And in the fast-paced world of executives we work with, time
is a more challenging thing to wrench free than just about anything else.
"Oh, great; now I have to add acknowledging people to my already full
day."

But acknowledgment doesn't have to take any longer than a passing
comment or handshake, visiting people at their workstations, or simply
adding a little bit to your existing executive routine. Consider this example
from one of our shared workplaces. In this particular global business unit,
each weekly senior leadership meeting started with a round of feedback
to recognize those who had gone above and beyond, followed up with a
short card signed by the president. Picture the pleasantly surprised faces
when these notes of thanks—often accompanied by a little treat—were
hand-delivered to employees!

In essence, we're talking about simple lessons learned during child-
hood. Remember how saying "please" and "thank-you" were drummed
into you? You know, we often wonder why as adults we come to believe
these fundamental niceties are no longer necessary, when they can make
all the difference in the world to those around us.

If we look carefully within, as Mary Kay Ash, founder of Mary Kay
Cosmetics, once commented, we will find each of us wears an invisible
sign around our necks that reads, "Make me feel special." Now, we may
not admit we have this need; we may be unaware of it. Still, part of you
lights up when you are made to feel important, even if you have difficulty
showing it or accepting it.

If you're already regularly acknowledging others in this way, take it to the next level: try not only recognizing accomplishments but also celebrating people for *who* they are. Recognize their important events and milestones while also discovering who they are, and what makes them tick. Rather than seeing people simply as human "doings," great leaders draw out others' qualities as human "beings" and make everyone feel valuable in their own right.

As emphasized in our coaching practice, the skill of acknowledging others is different from complimenting them. That is not normally a problem for capable leaders. In fact, CEOs often have an infusion of charm and social grace that makes it quite easy for them to give compliments: "I admired your work on that project." Acknowledgment is different and almost always begins with the word "you." Acknowledgment speaks to the unique talents a person possesses or brought to bear on an assignment that actively contributed to the result, not to the result itself: "You were so dedicated and persevering in the way you brought that project to completion."

Anyone can recognize the difference in how they would feel if they were the intended recipient of these two very different phrases. Most of us would politely acknowledge the first comment, and we may or may not see it as genuine. The second can only be taken personally—and it is intended to be. It is crafted to send a deliberate message that targets what another individual did with no intent other than to acknowledge what you noticed and appreciated about them.

To kick up your level of acknowledgment a final notch, we're inviting you to recognize results and effort for both successes and failures. This scares many of our CEOs when we first propose the idea, but it shouldn't. To be clear, we said nothing about demonstrating undue patience with mediocrity. No one is more aware than we of the disastrous consequences of failing to make plans—especially in companies whose stocks are traded on public markets. In fact, we are hired by companies all over the world to avoid this very outcome. We have a lot riding on the principle of making sure that our efforts deliver the promised results, since our livelihood depends on it! We understand that fear and performance anxiety.

So what do we mean? On this one, we're standing by our principle of a balanced view of performance that says if an organization aspires to balance risk and reward, this implies some level of potential to fail. In fact, a project portfolio that only ever succeeds might suggest (as we noted earlier) a lack of aggressive but informed risk-taking to ensure the organization remains at the forefront of what it does.

This means making sure that we don't penalize the talented teams who valiantly try but sometimes fail. Or whose projects are canceled because, as our strategy and competitive environment change, so do our priorities. Sometimes we stop a project rather than seeing it through because it's the right thing to do. Yet the project team worked as hard—maybe even harder—as those on successful projects, and they should be seen as contributors, not failures. Most internal recognition programs do not permit recognition of failure. Can you recognize the hard work and energy that went into such a project without feeling like you are endorsing a failure? Would you consider writing a note to each team member acknowledging how they probably feel but making sure you recognize the effort they expended trying to make it work?

Good leaders need to acknowledge people at all levels, all the time—personally, for who they are, and professionally, for what they do. If this does not come naturally to you, work on it until it becomes a habit motivated by your desire to succeed as a leader. As we have seen when coaching leaders, it is much better to be motivated to succeed than to be successfully motivated only after failing—so start today!

Activate

Let's bridge from our second to third executive "A" by pulling another unbelievable how-not-to case study lesson from the antics of the project management team leader, Marvin, you were introduced to earlier. It would be pretty obvious to anyone that Marvin was not going to be acknowledging others naturally. And it might even be true that in his organization there was no expectation set to do so, and that maybe this was part of the problem they were facing. This brings us to the role of the CEO as the *activator* of change.

Had one of us been the CEO, and on the basis of such anomalous turnover, we would have been asking some important questions about Marvin's organization and his base leadership skills. This would have been in spite of the results that Martin's team was delivering. We would have proactively intuited the possibility of an underlying problem in the front-lines rather than waiting for the evidence of an exception to come to our attention. This is a change in behavior for many leaders, who are often taught to "lead by example" and "manage by exception."

So we ended up being asked to coach Marvin around his leadership style. Now, from our very first encounter, we were not really sure to what degree Marvin agreed to enter into this arrangement with an open and willing heart. It was clear the new CEO had mandated the coaching. He had done so for several direct reports, ostensibly as a result of the recent acquisition and his desire to "integrate" the management team into its new structure. We suspect the CEO was using the degree to which his team reacted to both the offer and process of coaching to make decisions about their futures. However, at the time, neither Marvin nor any of us knew if this was the case.

We began our coaching and, to give him a moment's benefit of doubt, Marvin began to practice what he thought was good leadership by coming up with a brilliant idea. To help put this into perspective, it might be useful to have some idea of how we engage with executives and senior professionals in coaching. As co-authors and colleagues, we have written widely on the use of a co-active model as adapted from Schon's body of work on becoming a "reflective practitioner." It looks like this:

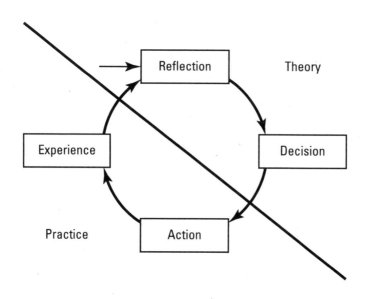

The Reflective Practitioner Model

This model presumes that people have what it takes to succeed (the inherent intelligence, skills and willingness to change) and that all we have to do as the coach is help them reflect on current practice and results, decide what they want to change, take action and reflect on the experience until it becomes a habit. Because it is not the intent of this book to summarize the behavioral psychology of effective counseling, let's just assume that early on we would have been in deep discussion with Marvin about how he asked questions and what results they were actually bringing him. This, in turn, would have touched on how to acknowledge the viewpoints and contributions of others.

Armed with the new knowledge that this was what he should do, Marvin launched a week-long campaign to let people know just how much they were valued and respected. He intended to provide treats at *everyone's* desk in the morning as they arrived for work, complete with pre-printed messages of his appreciation on them. Early on day one, the members of his relatively large team became aware of a hushed yet frenzied activity as Marvin, the HR manager, and his assistant all readied small cookies for distribution. Why the clandestine conversation among them?

It turns out Marvin had decided that only full-time project management employees on his team were to receive the treats, even though the department had many contractors and temporary employees whose workstations were located right next to these full-time employees. Picture the consternation over the strict instructions to give cookies only to full-time employees (despite the fact that Marvin's and certainly the department's budgets could have afforded cookies for all). You can imagine how this poorly executed gesture of "appreciation" only succeeded in making sure everyone knew everyone else's status, leaving many feeling that the value of their hard work had been dismissed.

Good intentions can be undone by insensitivity. A version of this story, called "The Cookie Incident," was featured in our first book, because of how well it makes this important point. What started as a novel idea became overshadowed by one manager's bureaucratic failure to keep his eye on the real ball—the desire to recognize people through a thoughtful yet inexpensive gesture. This was not the time to work out principles of the corporate recognition program. Marvin could have focused on speaking with each person he gave a cookie to, making sure they realized that he sincerely appreciated them and their efforts despite his gruff and unfriendly manner.

So what could we do to help Marvin? Well, as many executives reading this may agree, the starting point would have to be an awareness of what he was doing wrong and a willingness to try and fix it. Sometimes that is best accomplished with strong, skilled leaders like Marvin through one-on-one coaching, rather than training or development efforts. This method forces Marvins to reflect on and account for their actions. If we were his coaches in the aftermath of such an episode, we would move through the reflective practitioner model and use questions to get him to recognize what had happened. This approach is perceived as less invasive by those being coached and would maybe make Marvin less defensive about his behavior. Our questions for Marvin might look something like this:

1. What are you doing now that is or is not working?
2. What could you do differently that might get a better result?
3. What first step could you take to activate this change?
4. How will you know when the theory is working in practice?
5. Is the result worth the effort and should this practice become a habit?

As a CEO, we suggest you apply this technique in your own role as an activator of change within your organization. When you spot something that is not working and needs improvement, figure out how you can engage your organization in defining a solution it wants to try. This usually ends up as some kind of beta project, pilot or experiment to test the validity of the solution against the problem. Once the measurable results occur and there is certainty that this is the right thing to do, you get your organization fully engaged in doing it right. This helps avoid the "flavor of the month" syndrome where the CEO and senior executive team's attention spans seem limited to only starting new initiatives rather than seeing through the existing ones.

The reflective practitioner model provides everyone from the CEO on down with a more deliberate, focused way of delving into true organizational change. Not necessarily complex, the model serves as a simple reflection of how both individuals and organizations can be prompted to change through the power of self-diagnosis and purposeful action. This is where Marvin ultimately failed; his actions were purposeful only to the extent that he felt he *should* do them because they would *make things right* versus undertaking them with the certainty that he *wanted* to do them because they were *the right things to do*.

Given the inherent nature of the role of the CEO as the incubator of strategic organizational change, and the role of project managers in supporting the execution of organizational change, we hope this model can make a powerful contribution to helping those tasked with this outcome to succeed. And remember: "People can change people. People can change process. But process changes alone change nothing."

THE FORECAST FOR RELATIONSHIPS

A summary of everything we have just explored in the ABCs of Leadership might be most simply expressed as *leader-created relationships matter*. The forecast for the future is that they matter more than ever. Again, we refer to our central tenet concerning the critical role that effective personal leadership plays in cutting through the fog. As such, we are pointing to everyone's responsibility—executives through to each project team member, no matter their position—in deliberately crafting relationships based on who you are by way of your character, attitudes and beliefs.

For we find North American business particularly reinforces a point of view founded on getting the project work done first (sometimes at all costs); it is almost an unexpected "bonus" if people form good relationships in the process. At best, good relationships are an afterthought and certainly not a prerequisite to success.

Based on our many years of experience in a variety of consulting settings, we see the opposite as true. Just as profit ought to be the end result of doing everything right that leads up to a healthy bottom line, so too does task accomplishment naturally follow from focusing on relationship-building first.

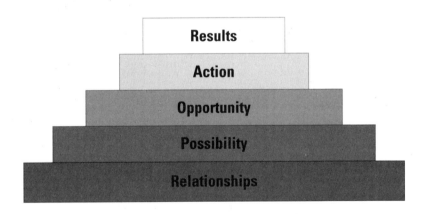

Relationships are the bedrock. To form this stable foundation requires *explicit* discussion about how we change the way we work. All too often, we take for granted that these critical ingredients will somehow magically come together. But ask yourself this: is it not true that poor communication and making assumptions creates a surefire recipe for project disaster? As we constantly repeat with our clients, you can either invest the time upfront in taking yourself through purposeful relationship-building dialogues, or you can leave this vital success factor to chance and experience the inevitable happenstance results.

The next building block, *possibility*, refers to creating ideas with others, based on the possibilities we can generate through our pooled wisdom, skills and experience. We move from building strong personal rapport (through relationships) into developing a set of "blue-sky" possibilities. As Einstein observed, "The problems of tomorrow cannot be solved at the same level of thinking that created them in the first place."

The building block of possibility is followed by *opportunity*. Only now do we move into strategy and plans. We translate the opportunities we see into a more tangible direction and business decisions, and then craft plans to execute on our strategy. Only when these cornerstones of relationship, possibility and opportunity are firmly in place can we ever move successfully into action and then on to achieving results.

It is critically important to notice that action and results follow from putting into place the other cornerstones, and not the other way around, thus demonstrating that how we are connected links directly to results. In the growing world of cross-cultural commerce, business leaders will no longer be allowed by other global "villagers" who value social engagement to impose their singular focus on the "what" (task) without also attending to the essential "how" (relationships), as was the case with an especially determined director we know.

Ina clearly believed teamwork would somehow evolve from having her project team focus on a high performance "stretch goal." At the time of one of her famous goal-setting exercises, team relationships were at best mediocre. That's because there was never time for the "fuzzy people stuff," seen by Ina as a huge waste of precious energy that ought to be devoted to the work at hand. As a result, team members did not appreciate what each person brought to the table or understand their roles and responsibilities. Nor did they truly respect Ina as their leader. However, Ina overlooked these realities in favor of basically ordering this disparate group of individuals to do their utmost to achieve the project goal. The idea was that by narrowing their field of vision on outcomes, they would achieve camaraderie.

While we can report that the project team accomplished their goal, we can also say it was strictly because their bonuses depended on it. Their sense of jubilation upon completion had nothing to do with improved team relationships, and everything to do with more money in their bank accounts. They reached their targets strictly as individuals, not as a cohesive unit. Interested only in doing the bare minimum to the get the job done, they created for the on-site project manager charged with bringing them together a task about as pleasant as a root canal!

This small story reinforces the fact that it is absolutely possible to achieve results without relationships. But we ask this: how much more powerful could the team's accomplishments have been had they also concentrated their efforts on developing true relationships? The project's outcomes would have been superlative, rather than merely "met expectations," had people been made to feel more important. If she had used the ABCs of leadership model to focus on her relationships, Ina may well have instilled an authentic team spirit, while attaining the solid results she needed for the business.

It all comes down to belief systems. Good leaders choose to value people as professionals with the inherent personal pride to get the task done, and done well. They are not cattle; we should not need to prod them. But to focus solely on the task is to focus on that which should be a given in performing the job.

By contrast, if leaders only direct their line of sight on people without an eye toward results, obviously companies would soon go out of business or donors would stop giving because they would see your organization as inefficient. In government, the corollary is that politicians will stop seeing value in your programs. We fully recognize the business realities! All we are really saying is that by rebalancing the weighting we give to task and people, we will bring about a healthy shift in disciplines such as project management, where the priority has been more squarely put on hard deliverables.

A PARTING REFLECTION

Great leaders strive to create deep belonging. They build relationships based on true partnership and connection, rather than mouthing the words "We're one big happy family, and people matter," while employees feel alone and forgotten in the overarching quest for results (where "only your most recent results count"). Is that what you really want your leadership legacy to be all about? The choice is ultimately yours.

And, given a desire to do things right while doing the right things, you can engage your organization in the changes required to implement PPM by choosing the right kind of leadership style. The kind that gets your employees or peers to recognize that the value of the work is best improved by making sure it's valuable—exactly what PPM can help you do. The rest will take care of itself, so long as you focus on the *what* and the *how* together!

## CHAPTER EIGHT

Sweeping Away the Fog in the Private Sector

This chapter explores successful PPM implementations from client organizations that have swept away project fog. As you will see, none of these firms ever intended to support project management malpractice! But they were. All the companies featured in this chapter are actually quite good at executing single, simple projects from beginning to end; their challenges arise when they execute multiple projects simultaneously. As the number of projects they undertook grew, they ended up unable to balance demand for resources among competing priorities. This is a fundamental problem for many organizations I work with, and often makes the C-suite believe they are becoming less effective at project management as an organization. The real problem is selecting too many projects in the first place.

The difference between a company that "gets it" and one that does not lies in the approach to this problem. Those that hope the problem will resolve itself do not get it; longer-term results will prove this. The winners, and those that set themselves apart in terms of improving the execution of truly strategic projects, ask themselves how they know they are proposing and selecting the best projects on which to focus their scarce resources. It is this important question that makes them ready to consider PPM as a next step in their project management practices. And this is the point at which I often get the call to help them.

When working with clients, I am often asked to summarize the benefits of PPM that a company can realize using my approach. I answer by identifying five areas:

1. Clarify your strategy.
2. Reduce the number of non-strategic project proposals.
3. Improve strategic project selection.

4. Increase overall resource utilization and decrease project risk.
5. Co-ordinate management of the project portfolio at the enterprise level.

CLARIFYING STRATEGY AND REDUCING NON-STRATEGIC PROJECTS

Project fog creeps in over time, but there are early warning signs that an organization can watch for. In the case studies that follow, executive team members often see the visible symptoms of project fog from their own perspective. But they can voice their concerns directly when interviewed about the challenges they face. Let's consider a few of these signs and symptoms of project fog from various executive perspectives.

The CEO is at the pinnacle of performance management, with a clear responsibility to the board to ensure that appropriate strategy is set and executed. A CEO (or general manager) will often see the first symptoms of project fog as a lack of specific connection between the projects being proposed for approval and the company's evolving strategic agenda. This leads to a lingering concern that the projects are perhaps not as strategic as previously thought. For instance, one CEO I work with put it this way: "Since I don't create the projects, how do I know that what I'm picking from is the only thing we can do to get us closer to where we need to be?"

He's not expressing a concern about how the organization manages projects; but rather is feeling constrained by the nature of the projects being proposed and selected—exactly what PPM addresses. Yet strategy is sometimes explained in ways that make it hard for those below the executives to recognize strategic potential projects. Employees propose anything and everything.

A CIO in a large financial services organization we worked with put it this way:

We never cancel any projects around here ... that would mean we failed. Instead, we redefine them, incorporate their scope into a new initiative or simply focus on new ones before the old ones are even finished ...

From the consulting work I do, this comment typifies the way many IT executives feel about project management efforts inside their companies (maybe because they are involved in just about every single project their company does!). But other executives often share this perspective of "we have too many projects." They report that as the number of projects increases, the organization eventually hits a wall and finds itself unable to replicate past project management success. Or that project after project is conceived and simply put onto existing resources, with an expectation that project work as well as business-as-usual work can somehow be handled optimally. This is unrealistic.

Yet when the CIO comes forward with resource requests to support estimated levels of project effort, the support of his or her peers often disappears; few grasp the complexity of managing this exploding list of projects. This is where strategic project portfolio management can make a difference.

When strategy is clearly articulated in measurable terms, employees have an opportunity to think not just in terms of projects that "make sense" but rather in terms of projects that "make a measurable difference." When projects are merely at the idea stage, they have an opportunity to compare the potential impact of their project against the gap between current and future performance levels to see just how powerful the idea is. As a result, only the most powerful project ideas tend to attract enough interest to be turned into actual project proposals.

You can see the benefit of this in an example from a major North American financial institution our team worked with. More and more of their retail banking operations were being performed using intermediating technologies (telephone; web). Of course, self-serve technologies are both more convenient for the customer and more cost-effective for the bank. As a result, the number of projects being proposed (and the associated human and capital costs to implement them) had become significant. Shortly after Darlene (not her real name) was appointed as a new EVP in charge of retail banking, she asked for a presentation from her new team on the projects currently underway across the retail division. She wanted to get a sense of what the major activities were because she had been appointed to implement a reduction in capital expenditures and an improvement in the return on equity in the division.

The starting point was a prioritized list of projects. She was shocked to learn that there were 137 individual projects or initiatives on the list. However, suspecting they could not all be of equal priority or value, Darlene had the team rank the list. Ranking the projects by priority proved difficult; they had previously only been grouped thematically into basic buckets and within that according to when they were to be completed. The chart shown in the meeting looked like this:

Customer Service	Process Efficiency	Cost Reductions
• CTI Script Renewal • Abandon Re-call Agent Release Patch • Auto-Push Online Survey Technology Install	• Expanding Call Monitoring Capability • Online postal code to address validation • Auto-fill fields for telephone loan application	• LD vendor RFP • Network/Switch Upgrade • "Banking 2.0" web project/online agent/ context-sensitive help

Similarly, while the bank itself had a relatively clear written strategy and some associated financial measures (such as targets for revenues; return on equity; bad debt recoveries and allowances), there was an insufficient amount of measurable detail at the retail bank level to ensure clarity. It was clear the bank wanted to be "a customer service leader in segments in which it chooses to compete for business"; it was not clear how this translated to their call center strategy. The call center was obsessed with measuring average time to answer (ATA) and call length and was convinced these were strategic measures. Yet it was not clear how these related to customer service or could improve the retail division's ROE. Meantime, good customer service was essentially measured on a combination of abandoned calls (customers who theoretically got frustrated and hung up before reaching an agent) and the odd customer survey administered by an external agency to a selection of customers who had used telephone banking in the past 60 days. Again, while these may have been important, it was not clear if they were strategic measures.

So Darlene was not surprised to see so many projects in the works. Basically, anything that seemed to "improve" customer satisfaction, process efficiency or reduce costs made the list. While offered with good intentions, the project list was no longer strategic—it was exhaustive. The fog had rolled in.

While businesses are always interested in talking about anything that can increase their ROI (or in this case the ROE), don't assume that simply because a project can be thought up, it is worth doing. There has to be a way to sort out the difference between a project that offers a financial advantage and one that offers a strategic advantage—with the two possibly (yet not necessarily) related.

We helped Darlene formulate three strategic questions for her call center team:

1. Where do customers get the best service, at the lowest cost to us?
2. How can we get more customers to interact with us that way?
3. Which projects (existing or new) can help us make that happen?

When she asked these questions, we saw a look of wide-eyed wonder as her team began to think about their business in a whole new way. This is what strategic clarity is all about—creating relevance that helps shape the context in which people work. After they had brainstormed the answers to these three questions, she asked them to develop a new strategic plan, specific to retail banking support services, to be used to guide decisions not only about projects but about everything they did. They developed several new goals, with accompanying measures, and were often able to link them to the core processes for which they were responsible. It was impressive work. Let me reproduce just one excellent example of that strategic clarity below. The goal read:

GOAL	MEASURES
"Our telephone banking facility is in the business of going out of business. Our objective is to expertly train customers on alternative self-serve systems, both telephone and web based, so that when they call for manual intervention the first time, it will be their last time."	% of reductions in manual transaction calls from customers measured monthly # of in-call customer conversions to self-serve or self-assisted technologies # of first chance queries to online auto-assist agents versus call center # of unresolved help desk calls requiring supervisor assistance % of customers rating help desk calls as "exceeding expectations" % of customers rating the online agent assistance feature as "invaluable"

If you are familiar with how call centers are typically managed, you probably find this example astounding. (Even if you are not, you can see how such clarity would instantly help in setting priorities.) When the team was done, the list of projects to be completed had dropped from the original 137 to a scant 43—and 14 of those were new initiatives, not even originally on the list!

More interesting to me, and more proof that PPM works: six of those projects cost the bank money in the short term but had the longer-term potential to switch customers from the telephone to the web and were now considered essential priorities. These types of projects were new. Darlene saw that her team was on track to make a strategic difference, be more selective about which projects it did, and reduce capital expenditures while improving long-term performance.

IMPROVING PROJECT SELECTION METHODS AND PROCESSES

Some of my earliest work on PPM began in February 2002 with a key presentation to the executive team of a very large financial services firm (with over $11 billion dollars of assets under management), thousands of employees and national brand recognition. This Canadian organization, a leader in its field, is highly profitable and well-run. The CEO and various members of the executive team are long-serving and have been cited in various publications and earned awards for their work. They were a "perfect client" in so many ways.

They already used a balanced scorecard and had a very clearly articulated strategy and associated measures. While the implementation was not perfect and there were some gaps and some areas where they had made their strategy map overly complicated, nobody could say they didn't have a clear strategy.

Of course, a firm of this size has a large number of projects underway at any one time. Their active projects ranged from several hundred thousand to several million dollars in terms of budget, and varied in timeline from several months to five years. Yet every year, they struggled with the perennial question: which projects to approve and why? In fact, this so bothered a particular VP that after she saw a presentation I did at the Conference Board of Canada she asked me to make the same presentation to her executive peers. In her own words: "There has to be a better way to select projects than our annual drag-them-out, knock 'em-down budget scrum." Even though their strategy was clear and measurable, as I stated in the opening chapters, this is proof of the fact that executive teams still tend to pick their projects on an other-than-strategic basis.

This illustrates a key point of my research about PPM: even if your overall strategy is clear, there is still a need to link it explicitly to your internal project selection process. Otherwise, you risk having a cumulative performance gap, which I define as the difference between what is expected (in terms of measurable strategic and financial gains) and what is selected (in terms of cumulative benefits of approved projects).

The approach we chose to start with was to generate executive buy-in about the need to make changes in the project selection process. To achieve this, I had to establish that there could be gaps worth addressing in their current practices: possibly the projects being selected were not always strategic and performance, while good, could be even better. We interviewed every executive at the beginning and end of the process. First, we asked them a series of questions on the current, "as is" state of project selection in the organization.

In presenting the results back to the client, I had agreed that their individual identifies would not be disclosed to their peers or the CEO. This allowed for frank, revealing interviews; we are reproducing these questions here in case you also wish to use them:

1. In your opinion, how are projects selected and prioritized today? Is this effective? If so, why? It not, why not?
2. What aspects of project management are working well today? Which are not working as well? What would you like to do about them?
3. Given what you have seen of the proposed PPM methodology, what aspects of this intended process seem of most value to you from your vantage point as an executive in the firm? Why?
4. What are the potential barriers or issues during implementation that you feel we have to watch out for? What can be done to limit these in advance to ensure we can successfully design and launch the new PPM process internally?
5. Is there anything else you wish to comment on at this time that you feel would be helpful or useful for me to know?

This approach has since become a feature in all of my consulting work with corporate clients, because once they understand the gap between their existing practices and a truly effective PPM methodology, there is usually agreement and support for improvements. Of course, this is what most project management professionals want for their clients or employers: to be given license to help them develop a strategic project selection method that both works and is actually strategic!

Perhaps using tools like interviewing key stakeholders to establish the gap will help you gain traction and interest in PPM within your organization. Again, to help you, here are the five follow-up questions I use once PPM is in place to establish if the methodology and process are working.

1. In your opinion, how are projects selected and prioritized now using PPM? Is this more or less effective than before? If so, why? It not, why not?
2. What aspects of project and portfolio management are working better today? Which are not working as well? What else can we do about that?

3. Given you now have experience with the PPM methodology internally, what aspects of this process seem of most value to you from your vantage point as an executive in the firm? Why?
4. Given what you expected when we began this process redesign, what turned out to be the biggest opportunity? Was it what you anticipated? What was the biggest barrier? Did we solve it successfully? Why or why not?
5. Is there anything else you wish to comment on at this time about the process we have just completed? Would you do it again? If so, why? If not, why not?

In this particular company's case, the executive team immediately saw the value of what was being proposed and the CEO lent his support to the endorsement of a project to be led by the firm's corporate project management office (CPMO). It is an essential point: to function properly, there can be no "standard process" for PPM. Each organization must take the standard process steps and associated measures and link them all to their unique strategy.

We also discovered during this case study that this level of process change is best accomplished using an internal process team but facilitated by an external PPM expert. This finding emerged after a couple of false starts where I used primarily consultant-driven teams, but with much less success. The PPM process is too strategic to a company's destiny to be left in the exclusive hands of consultants. Yet companies need expert input to stimulate their awareness of the need to change.

For example, in this company there was one function (IT) that was initially facilitated by an employee, but we discovered that the work done there did not go nearly deep enough, or sufficiently probe gaps in process, to draw the necessary conclusions about effectiveness. Often internal facilitators, especially in organizations undergoing change, cannot be sufficiently direct and blunt to be effective, and it may be an unfair for executives to put employees in this situation. We genuinely recommend that CEOs and project managers considering proposals for PPM process projects look at the prudent use of external facilitators in their initial diagnosis of any gaps.

In addition to looking at their current processes, I conducted a series of structured interviews with key executives and other managers inside the firm. This resulted in a final three-phase implementation plan for a new PPM process:

- Phase I: Process design
- Phase II: Implementation of new core PPM process
- Phase III: Implementation of supporting tools (capacity planning; estimating and project activation; tracking and control, including desktop access to project status)

It's a lengthy process. We initially estimated that all three phases would take between nine and 12 months of work, with a part-time consulting team of two or three to support the internal process work: design expertise, stakeholder interviews, facilitation and training support, creating supporting tools, and so forth. In the end, we completed the project in about 15 months.

An essential component in the first phase was to convert their clearly stated strategy and associated measures into a project scoring model. The corporate project management office (CPMO) and I began this task by embedding their strategic measures into the base template used within the organization to propose and select projects. This work was precipitated using a straw model to provoke the team's thinking on how to redesign their existing and quite workable project proposal tool. It needed to integrate strategic project scoring early on, and would allow them to make relative comparisons between projects on the basis of measurable project outcomes. (This is what I mean when I note the need to explicitly connect strategy to specific elements of the project selection process.)

For the next three months, consultations during workshops and training sessions explained the new project scoring and project submission process design. A cross-functional team was assembled that included representatives from all groups impacted by the new process design, and the role of a "champion" for each area was assigned to a leader/manager from within that group. All of these tactics are taken directly from effective change management practices, and their use is not unique to this project-management related example. The champions took center stage at these workshops, along with the CPMO staff and our team, in order to help the new process gain credibility with important stakeholder groups.

The feedback received affected the process design dynamically. Improvements suggested by those on the "front edges" of this new process iteratively improved its ultimate, final design, while still respecting the essential elements of the generic PPM framework I presented earlier in the book.

However, one of the most important things we validated was the shift towards executives being able to select projects for their strategic importance rather than on their originating department. Nine out of 12 executives (75%) indicated during this period that the process had improved the project selection process of the organization, and 100% indicated that there was now an emphasis on "on-strategy" selection of projects. Additionally, 80% (10 out of 12) of the executives felt the process was a significant improvement on internal practices, and they found this change sustainable. That same 80% felt there would be a significant improvement in the quality of projects proposed and an improvement on the use of capital within the company. These are the kinds of important behavioral shifts that PPM can create.

With the core process design completed, and Phase II beginning to unfold, the corporate PMO undertook responsibility for identifying other core processes inside the company that would be impacted by the new PPM process. Once a co-located process was identified as requiring a change, they began to work with the appropriate stakeholder groups to determine how to link the new PPM process design to processes owned by others.

This effort ultimately reached across the company into such representative areas as Finance (related to how project business cases are assembled, documented and tracked), Human Resources (for changes in how project resources are identified, costed, and hired), Internal Audit (especially as it relates to embedding new strategic measures into existing reporting mechanisms for post-project completions and benefits realization) and IT (especially relating to software development methodologies, estimating processes, resource assignments, time tracking). The executive interviews affirmed the worth of the process—it reduced the number of projects and ensured they were aligned to strategy. However, many commented on the complexity and cost of the process, and they mentioned the amount of support needed from both Finance and IT to make the process work. For instance, one executive commented:

> This was a substantial investment of my and my colleagues' time that will have to result in cost savings or improved project outcomes to justify . . . it's a good process—solid and makes sense—but our internal processes such as IT estimating or capital budgeting are going to have to also improve if we are to benefit completely.

You cannot assume that implementing a fully functional PPM process will only involve your project management processes. PPM inter-relates to a large number of internal company processes. Recognizing the level of process integration required has since been identified as a crucial element of success. In PPM engagements I do with clients, we discuss this; such integration is a clear "better practice."

As Phases I and II were mostly completed during the course of that year, the new PPM capability was now functional and operational across the firm. They didn't really need consulting support any longer. However, during the process, executives anecdotally reported a significant degree of satisfaction around the value and benefits of the new process, and particularly mentioned its ability to link programs to strategy. As I write this, the methodology is still in use within this company. So it's worth asking: what made this implementation successful? How could you replicate it if you were trying to accomplish the same result?

One aspect of success that may or may not be replicable was that the senior leadership team (SLT) were generally a highly motivated and successful group of executives who were willing to change the project selection process. While sparring and debate had been common in previous annual planning sessions (where projects were proposed and selected), the team knew that was not optimal. It had been political. Sales and marketing, in particular, seemed to have the upper hand, getting projects approved through the threat of declining revenues.

Using PPM, executives compare projects against pre-determined strategic criteria and measures, which eliminates much of this behavior. This helps them be more strategic. While this is a positive outcome and was generally positively received as a beneficial outcome of PPM, it may make executive teams who are less secure or more attached to personal power feel threatened. Be aware of this potential dynamic.

However, support for the process was high, and most executives were in favor of continuing to use PPM as the primary method of selecting and approving projects. Many commented they would do this again at another organization or would recommend it to peers as valuable.

TRAINING AS A TOOL OF STRATEGIC CHANGE

Another important thing we learned from this company was the value of taking time to train employees on the new process and its supporting tools. These sessions were a significant investment of time and effort, held in small groups of 20 to 30 participants and lasting three hours. However, these sessions were used not only to train participants on the new process but to also get them engaged in defining strategy at a measurable level for their own function or department by trying to imagine and define a strategic project proposal. This helped with buy-in and with strategic clarity.

The approaches used in the training sessions were generally highly rated, getting a score of 4.6 out of 5 in terms of satisfaction. They were also appreciated for their overall conduct and effectiveness (i.e. interest, pace, quality of materials and effectiveness), getting a score of 4.4 out of 5. This indicates participants were generally positive about the investment of time and effort required to attend the training.

However, the most common concern expressed by participants in their write-in comments was the perception of the process as "administratively cumbersome." Both the consulting team and executives saw this as a reflection of the fact that the firm had no specific standardized process for project submission prior to implementing PPM. Projects were essentially summarized by the sponsor, and support at the executive level was confirmed during the annual budget process. Here is a selection of their write-in comments from the training sessions:

- "Watch out for 'methodology madness' and putting way too much process in place for accomplishing a simple task . . . "
- "My impression from the session is that there is a disconnect between the business and IT: the business does not appear to understand the value and benefit of a methodology and just wants results, fast!"
- "Once we find we can't get the projects approved at the speed we are used to, will we stick to this or change it all over again?"
- "While this is a good system to help us select projects, I still think some executives will be able to get their own way and get their 'pet projects' approved or else they won't support it."

Over time, I have realized that this is a common starting point and may reflect the reality inside your organization today as well. There's a real risk that a standardized, detailed process of project submission will be seen as too "complicated"—unless you convincingly pre-sell the benefits. Even though we did so in this firm with its executives, we did not perhaps do it sufficiently with employees to establish the rationale and need for the new process before we started training them on it.

Participants commented on the very significant departure this represented culturally inside the corporation—moving from executive sponsorship as the primary asset for getting a project approved to shifting and sorting projects on a strategically prioritized basis. Some participants were skeptical about just how committed the senior executive team was to this change. This fear of the unknown was evident in the comments on the participation forms:

- "It sounds like the most challenging part of this process will be changing our corporate culture to respect process and not politics."
- "Get a communication consultant assigned to this project team so that communication is consistent, credible, and clear . . . and targeted so that those who need to support the new process are in the loop early."
- "While the training was great, I will wait to see this process in action before deciding if it will work for us."

Again, the implementation team noted these were offset by generally positive comments in other areas of the form about the firm being "on track" and "finally getting around to doing something that should have been done years ago," and exhortations to "carry on" and "keep us moving forward." So concern is balanced with optimism that the process will work and add value within the firm. In terms of replicating success and learning from this example, any firm undertaking PPM should be aware that vigilant employees will see if the executive team "sticks to their knitting" and will be watching for signs of visible and consistent support of the process.

Finally, when asked to comment directly on the merits of the new methodology (on which they had just been trained in detail, including the use of new forms, online tools and strategic measures), the range of comments was pretty consistently positive. These participants included executives, managers, supervisors and frontline employees from various branches and departments. Some had extensive project management experience; some had very little. Some had been involved in the beta process work and others may have only been seeing the process for the first time. Here's some of what they said:

- "We are getting to a good place . . . keep going."
- "I like what is being done . . . it is clear and simplifies project planning."
- "We can now say 'no'—the new PPM process gives us parameters to determine when a project should not be done—just make sure it's OK to exercise this option with executives!"
- "Overall, process looks very good. There will be some short-term pain but long-term benefits for sure . . . this is really going to help us cut down on useless work on projects that don't go anywhere half the time."

- "I think the process is valid. I look forward to being more involved in this project."
- "This is a lot to learn but the process design seems sound. I'll use it . . ."

These findings show strong support for the proposed PPM process framework and evidence of a beneficial impact on the firm. This is significant and substantiates the problem—the need for a strategically oriented project scoring tool and its ability to help the firm prioritize project work and manage their project portfolio for strategic outcomes—and a sense that our method can validly address this.

INTEGRATING RESOURCE MANAGEMENT INTO PPM

Beyond the obvious benefits in selecting strategic projects, PPM helps companies make better use of their scarce resources and improve their understanding of risk. However, the importance of scoring models is less about the actual scores and scoring models and much more about the ability of PPM to inform decision-making. It vastly improves executive understanding of the trade-offs involved in accepting various levels of project work inside their organization.

For example, one of my clients is World Travel Protection, which is a worldwide emergency services provider and a division of Zurich Insurance. Theirs is another example of a long-serving executive team that works well together with a deep sense of both collaboration and commitment to doing what's right and a clear understanding of their business and industry. In fact, they became aware of what I was doing when the CFO attended a conference at which I was doing a workshop on PPM and immediately resonated with one of the questions in that presentation: "No matter how hard you run, do you find yourself unable to keep up?"

Like many high-performing organizations (and WTP had undergone considerable market- and cost-driven challenges in the previous few years), while there may be an ability to identify, propose and even select strategic projects, often I find there is no real connection between this process and the available project resources. This was the case at WTP and was a vital step in the PPM process implementation from which we can learn some valuable lessons.

Without a specific understanding of where the resources are going to come from that will execute a project, a company's well-intentioned executive team can overcommit and actually increase project risk, simply by ignoring this question. Yet, with foresight and thought, most executive teams could determine *alternative resourcing strategies* (ARS) that creatively and properly address those issues. Let's look at why this happens.

Given their corporate history, at WTP a premium was put on sustainable earnings and the accompanying cash dividend paid up to their corporate parent. While this is an important consideration, what had started to happen was that the executive team mostly selected and approved projects only with internal resources attached to them. This was culturally imbalanced by a perception that these projects were "cheaper" to do. Eventually there were simply too many projects; internal resources were unable to keep up. One of the project managers inside the company said, "Sometimes we're not all that rational about this . . . we simply keep working and hope we can get it all done. Often we do. But at what cost, and can we keep this up?"

This was not a strategic approach to resource allocation and the CEO knew it. In a similar way to the gains to be made by measuring your strategy, an executive team should have an ability to look at its internal resource capacity and capability for projects and allocate this *strategically* to the top priorities. This can be done using a simple resource management system that determines what kind of typical resources are used on projects. At WTP, for instance, it was clear that many projects involved call center team members because of the importance of excellent quality customer care in their business. So they determined that there were four representative kinds of resources for any call center project: an agent, a staff member in claims, a trainer, an account manager or an executive in charge of the call center. With these categories determined, an average cost for each of these categories can be calculated (based either on typical costs or, in the case of a named resource, a specific cost), along with an average availability of that resource to do projects. This takes into account things such as discretionary time, vacation and holidays. All projects proposed and approved are then tallied; you can quickly determine where there may be resource gaps. (See the following page.)

Almost all of the work at WTP was done using basic EXCEL spreadsheets like the one on the following page. They were developed internally by one of their extraordinary project managers. Heather was able to work with us to identify formats and formulas to present information like this to her executive team enabling them to make better project and selection decisions. This is a good example of the value of PMs being at the executive table . . .

What is interesting about this specific example is that it shows a typical pattern for my clients: because executives have more control over their time

WTP

Resource/Day available	Maria	Days/Months	Acct Mgmt	Days/Months	Trng/Qual	Days/Months	Clms	Days/Months	Csv	Days/Months
		87.0		94.3				9.4		4.7
# of Resources in the Pool	1		4		2		9		70	
% of Discretionary Time	80%		75%		10%		4%		2%	
Available Project Days		188		176.25		23.5				
TOTAL PROJECT DEMANDS	101		82		190		14		21	
AVG # days/month = 22	22		22		22		22		22	
# person months of effort	4.59		3.73		8.64		0.64		0.95	
# days/year = 235	8.5		8.0		1.1		0.4		0.2	
(265 - 10 stat - 15 vacation - 5 flex)										
RESOURCE GAPS:	87		94.25		-166.5		-4.6		-16.3	
Excess (-shortage) months	4.0		4.3		(7.6)		(0.2)		(0.7)	

(with an expectation that they know how best to use that time to get their jobs done!), they often have more discretionary time available to do project work. However, their time is also the most expensive and most valuable, and what is often really required for projects to deliver the best "bang for their buck" is to assign more junior staff to them to actually do the work. As in the WTP example, though, that is where the resource crunch most often arises. I call this "Titanic Syndrome"—we see the tip of the iceberg, but down below is where the damage happens. While an executive may have sufficient time to do project work, does everyone else? And do your project returns consider the actual cost of this "free" time in relation to alternatives? And do we really have the right resources working on the right strategic projects to ensure they succeed?

As the WTP executive team began to see the resource issues more clearly, they were faced with some interesting alternative decisions:

1. Bring the number of projects in line with available resources.
2. Increase the resources available to do projects internally through hiring.
3. Adjust the resources by outsourcing or acquiring alternative external resources for project-related work.

Once they understood this, the reluctance to consider alternative approaches (even if those cost money!) disappeared, and some learning took place about how best to structure project work internally to maximize return while minimizing costs. This does not mean defaulting to always using internal resources; in fact, it may show there are times when critical, strategic projects demand that you hire either more resources or use externally contracted resources to accomplish the work. Do consider resources as part of the project selection process; but make sure the facts are on the table when executive teams make these important decisions.

SUMMARY OF RESULTS

What benefits can project managers and executives expect from the hard work required to implement an effective PPM process? Well, based on the case studies cited above, here is a brief list:

- The number of total projects approved likely declines.
- Capital expenditures on projects go down; returns go up.
- You can align risk and return based on your own organization's history of project success rather than on generic factors provided by someone else.

- Your project selection process will be more strategically aligned and you can "build in" success as a consequence.
- Executives and project managers will likely find themselves more engaged in process and less engaged in politics as the major driver of project selection.
- Other important business processes, such as your project resource allocation and management system, will take on new importance and have more of an impact on project selection decision-making.
- Staff will find themselves more able to focus on the important rather than the urgent—and projects will have greater certainty of being completed on time and on budget.

## CHAPTER NINE

Two Case Studies: Clearing Up the Fog in the Public Sector

Is it possible to clear up the fog in the public sector? As I mentioned earlier in the book, the policy-driven complexities of this sector make it a particular challenge. Yes, this is one of the reasons my colleagues and I so enjoy consulting to this unique sector.

Most public sector and not-for-profit organizations have multiple stakeholders and complex planning strategies; these create different tensions in project selection. Similarly, the absence of a single, defining motive (i.e., making a profit) means the organization's purpose is multifaceted and not easily summarized. However, effective project portfolio management should still be a fundamental goal for these organizations because of the benefits it can offer them in their more complex environment.

Let's consider this in relation to a typical board of education, which is one of the examples to follow in this chapter.

Any board's leadership team should focus on the primacy of students and student success. Yet they must also consider other important stakeholders. The board considers parents and their perceptions of quality of their child's education; the opinions of the system funders (in Canada, the local municipal ratepayer); the morale and contributions of teachers and administrators; the bargaining positions and contracts with any relevant unions; and the policy and regulatory environment imposed on boards of education generally by their provincial ministry of education. This convergence of interests, complex to manage in and around, is common in the public sector.

As we saw in the previous chapter's case studies, the approach we used to implement a strategic PPM framework was successful in the private sector, and five major benefits were seen in those case studies:

1. Clarifying strategy through measurement
2. A reduction in the number of non-strategic project proposals being made
3. An ability to improve the selection of strategy-focused projects by connecting project outputs to measurable impact on results
4. An improvement in overall resource utilization and productivity through improved central project activation processes
5. Co-ordinated management of the project portfolio at the enterprise level.

After this initial success, I was keen to see if the same PPM process framework that had emerged from our work in a for-profit setting would have similar value and achieve similar results in the public sector.

CASE STUDY CANDIDATE 1

In 2002 the Peel District School Board was already an acknowledged leader in their field, with a reputation for innovation in education. The strong leadership team, led by Director of Education Jim Grieve and Associate Directors Judith Nyman and Wayne McNally, and supported by an outstanding group of administrators and superintendents, struggled with how to keep their school system moving forward while responding to concerns about overwork, too many centrally sponsored projects and the lack of a strategic focus.

The Peel District School Board (PDSB) is one of the largest of its kind in Canada, comprising more than 200 individual school sites, over 11,000 employees and more than 130,000 students across a large territory just outside of Toronto, Ontario.

In Canada, as is often the case elsewhere, school boards are funded by a combination of grants from the provincial government and taxes imposed on city residents. The system is governed by a group of elected trustees. However, the day-to-day management of the organization is ultimately the responsibility of a director of education and a group of academic and corporate officers. The management teams are mostly drawn from the ranks of certified professionals, such as educators (typically former teachers), accountants, planners or human resources professionals. Paralleling the private sector, the organization defines their management structure to separate the operational arm (divided into regions called "Families of Schools" on a geographic basis) from board-wide, centralized support functions such as Finance, Human Resources, Accommodation and Planning, and Curriculum and Instruction. These common functions are housed at a central board office. This structure is similar to the private sector concept of "head office" and "the field." However, when applied in the public sector, it does not have necessarily have efficiency or profitability as its central tenet; the concerns are more about consistent policy compliance and local service levels and responsiveness to communities.

The "product" in this case is the formal and informal education of students to a level required in a prescribed provincial curriculum and validated to some extent by standardized testing and other practices such as student retention and graduation rates. Internally, this effort is referred to as "achieving student success." On the surface, the mission of teaching kids seems easy, and the optimal strategies might seem intuitively clear. However, these strategies are conceptual in nature. Their effectiveness is hard to measure, given their reliance on social policy objectives and outcomes, and given the nature of the multiple stakeholders that must interact to make it happen. This last point illustrates exactly the weakness of existing PPM methodology for applications like this: there is virtually nothing related to financial efficiency or rates of return in accomplishing "student success," but, clearly, there is still an expectation of financial efficiency and appropriate stewardship of taxpayer funds.

So there is often no defining underlying strategy to marshal all stakeholders in the same direction. The purpose everyone agrees on—educate students—can't prevent a struggle over how to accomplish that goal.

The core of this strategy relates to more balanced outcomes such as effective curriculum and teaching practices; the use of technology in the classroom; reducing teacher absenteeism; and the state of relations in the school among staff, students and the surrounding community. These dimensions of performance are far more central to students' achieving their full potential than any financial objective, but they are also harder to measure, a situation that makes it tricky to select the most strategic projects for this organization to pursue. Just what is "student success," and how do we measure it?

So, at the outset of the case study, the organization was struggling with the issue of how to make strategy more measurable (apart from reporting standardized test scores, as mandated by the province—which seemed too narrow to capture the essence of everything the board does to promote individual student success). They wanted to determine the specific, identifiable drivers of student success, to focus the organization's resources and efforts on the most strategic projects. This meant establishing "cause and effect" relationships between their system-wide efforts and the organization's overall results.

STEP 1: CLARIFYING THE STRATEGY

For many years the Peel school board had too many projects and initiatives launched centrally into the entire system. This is a classic symptom of project fog and should be of concern to a leader when it is discovered. The overwhelming flow of data, requirements and requisite activity created challenges at the local school level. School principals and vice-principals found it increasingly difficult to cope: "If I actually did everything that everyone wanted me to do, the job wouldn't be doable." Since each individual principal decided what to work on; this defined the work that actually got done at the local level and became their "pseudo-strategy." In that moment, it was the strategy. Even if projects were mandated centrally, they would fail to achieve their planned impact if they were not consistently implemented.

While every single project or initiative that was proposed and undertaken was done with the best of intentions, no strict measures were in place to define "student success." So how were conclusions being drawn? One school administrator said, "I choose projects that I think will be best for my students and my school." When questioned about criteria he used to define "best," fuzzy logic and unclear answers emerged. As experienced teachers and dedicated administrators, they often used their intuition and instinct to make these decisions.

Yet, this was the real question for the organization: From among all of these potential projects in place today and being newly proposed, which ones are absolutely essential to student success, and which ones were spurious or inconsequential?

To answer this focused question, the board would need a stricter regime of measurement.

The executive team made a bold decision in 2002, based on a proposal from the associate director and supported by the director of education, to use the balanced scorecard methodology. They turned to a well known local consultant to help them and after a substantial effort by the senior leadership team, the "Report Card for Student Success" was created. This turned out to be a very strategic choice by the Associate Director to help them address their strategic challenges. She used a five-step process (shown below) that can help any team understand what it takes to accomplish this work, and address the dual challenges of reducing effort while improving results:

The Five-Phase Implementation Plan

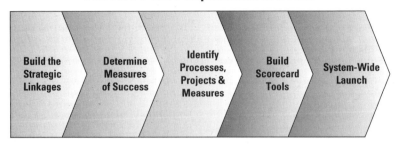

As they worked through the first few phases of this implementation plan with the other consultant, the focus was on following the steps originally defined by Kaplan & Norton when an organization tries to create a balanced scorecard. A representation of the results of the team's work over several months of intense meetings is shown below in the form of a "strategy wheel," with student success clearly in the center:

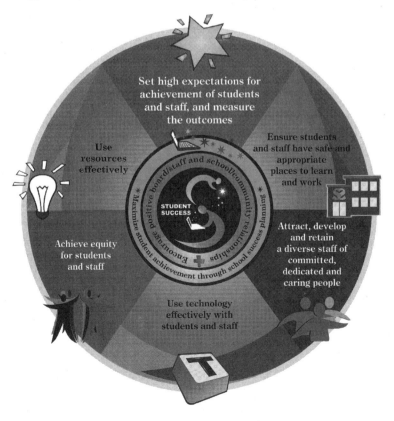

For each goal, clearly defined measures and targets were taken from the initial work done to create a strategy map that showed strategic pathways for achieving student success. The entire package created that year helped, for the first time in their history, to define just exactly how they were going to measure "student success." This clarity of purpose was essential to the results that followed and communication tools were used to secure board-wide commitment to it. The director said to me during our later work together, "We now have the ability to consolidate our effort and focus on our purpose like never before . . . this outcome will make a difference for what we do in Peel for years to come."

To those of you familiar with more traditional balanced scorecards, this wheel may not seem to follow the normal approach. For instance, there is a need to adjust some of the language around words like "clients" or "customers" and replace it with "students" or "stakeholders." In fact, when I joined the project this was one of the first things I was asked. In my opinion, none of these labels changed the intent of the balanced performance methodology itself and I would encourage organizations to undertake this review and ensure that the approach they are using to measure performance resonates within the organization and its stakeholders. Otherwise, there is a risk the use of inappropriate language that does not resonate with your frontline staff will cause the strategy to remain confused or obscure, defeating the whole purpose of the strategy clarification effort.

If you map the four traditional perspectives of a private sector BSC (financial, customers, process and organizational learning) over top of Peel's strategy map, we can see how the "strategy wheel" works in practice. Remember, it is not the use of the BSC per se that is so important here. Rather, it is the concept of creating a balanced set of measures that represent your underlying organizational structure. For those of you reading this who already have a balanced scorecard in some form, this may offer a solid starting point. In fact, if you already have a BSC then implementing project and process portfolio management is much easier. However, if the balanced scorecard is unlikely to be welcome in your organization (perhaps because of a past failed implementation or a concern about the complexity of these kinds of approaches), then a solid default position is to simply attempt to measure your organization's strategy using KPI's across similar dimensions but renamed or renumbered to suit. The essential point is simple: you must be able to reduce your strategy to balanced measurable terms that include more than financial goals and objectives.

PEEL DISTRICT SCHOOL BOARD

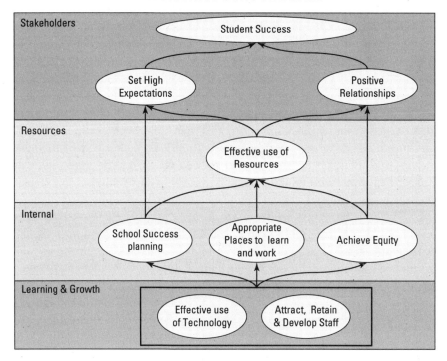

During this phase of identifying strategic linkages and individual process and proxy measures, the Peel strategic planning team also created a list of projects currently associated with each element of the strategy map, to determine if a balanced effort against strategic outcomes was currently present. When this step was completed, the PDSB leadership team had assembled a 50-page list of "strategic projects & initiatives" and an overwhelming sense that they needed to prioritize and focus much more if they were going to be successful at executing their strategy going forward. They had arrived at the next step in the generic PPM process.

STEP 2: CONSOLIDATING EFFORT

Once the strategic measures are clear, these same measures can be used to assess the "measurable strategic contribution" of any current or proposed project to establish its relative strategic priority. To accomplish this, the PDSB focused on identifying and stating the absolute contribution of any project to the targeted performance level of one or more of their strategic measures. Would the planned outcomes of the projects directly, indirectly or only weakly affect measurable performance?

Obviously, "strategic" projects touched more of the targeted measures more directly; they had the highest potential for contributing to student success overall. An example of how expected contributions and measures from the Report Card for Student Success were assessed and balanced as a portfolio is shown on next page.

The board attempted to balance the project portfolio across a range of diverse goals (shown across the top) and then link specific projects to impacting those goals (the green boxes). As the PPM process moved through its internal maturity curve, this was refined to include an assessment of strength of contribution and risk on a project-by-project basis that began to have an even bigger impact.

The benefit of this strategic work was not isolated just to projects in this particular case study. We also used it to identify a number of key internal processes that had the potential to be major contributors to performance and assigned them "Champions" who were similar in role to a project sponsor. These combined efforts (focusing on strategic projects and process improvements) would mostly determine the organization's ability to reach its strategic goals. Finally, as they realized that there were too many projects underway and that the organization could not possibly focus on that many, the leadership team culled through the huge project inventory. They assessed which projects contributed, and to what extent ("strength of contribution") and then made the bold decision to keep only the highest-performing projects in the portfolio.

As Judith Nyman said, "We went from so many projects we couldn't track them all to a small set of carefully selected projects that we knew from our analysis had the potential for high impact." This is exactly what a well-run PPM process should do for you—it should promote confidence that the right work is being done that will create the right results.

It wasn't easy. And it took about two years to get it all done. It involved significant group discussion in the director's office and frequent meetings with centrally-based and field superintendents and controllers about how project contributions were being assessed. But finally the approach began to take hold and make sense. This was the outcome of this intense effort:

- the Report Card for Student Success, published system-wide
- a substantial investment in training workshops, enabling system-wide communication of the process, the outcomes and future impact of the RCSS and then tying it to a local school success planning process for each site to ensure a focused effort.

Peel District School Board Project Grid

Objectives	Stakeholder		Fin.	Internal			Learning	
	High Expectations	Positive Relationships	Effectively Use Resources	School Success Planning	Appropriate Places to Learn & Work	Achieve Equity	Effectively Use IT	Attract, Retain & Develop Staff
#	S1	S2	F1	I1	I2	I3	L1	L2
Wt.	10	15	10	10	10	15	10	20
Projects								
Measuring Success Data Project								
✦ Develop and implement the School Success Planning Web site								■
✦ Develop a data mart				■				
✦ Develop data standards for organizational decision making								
Align and Consolidate Program Delivery			■					
Increase K-12 Literacy and Numeracy Skills								
Establish Org. Wellness Plan	■							
Implement the recommendations of Streamlining Committee		■						
Recognition Plan for Students and Staff		■						
Conduct survey to enhance School Council Effectiveness								
Integrated System Implementation of "The Future We Want"						■		
"Pathways" Resource Allocation Project								
Installation of Security Cameras in Secondary Schools					■			
Implement "Warning Signs"								
Admin. pre. and post input to School Planning & Construction								

- the publication of a list of key projects and processes associated with each measure and establishing a "champions group" of key leaders to individually take responsibility for each of the goals on the strategy wheel
- investments made to design automated data collection and dissemination tools (in the form of an RCSS website) to encourage data-driven decision-making and to report on measurable results against targets.

So far, the results have been so successful that even as of the editing of this book in late 2007, the PDSB was still using the original RCSS and various aspects of the generic PPM process to continue to drive its annual school success planning and project portfolio management processes internally.

STEP 3: REINFORCING WHAT WORKS

When you use dynamic strategy measures rather than purely relative project-based measures, our PPM framework allows an organization in either the public or private sectors to make informed rather than incidental decisions about which projects should be chosen. This tends to inspire informed action and data-driven decision-making throughout the organization.

In particular for the Peel District School Board, the PPM effort was further intended to reduce the volume and flow of projects emanating from the organization's central office from the "many" to the "few" in order to increase organizational focus and increase the likelihood of maximizing strategic results. Especially in the public sector, this is a critical goal that often determines how successfully an organization is performing. Jim Grieve, Judith Nyman and Wayne McNally have proven the value of this approach in an educational setting through their stamina and determination to stick with it in the last five years.

They were able to reduce the level of system-wide project activity by over 40% as a result of having a rigorous measurement system attached to the definitions of "student success," a formerly positive but ambiguous statement of intent. Anecdotally, system-wide reports during subsequent school years of operation noted a significant reduction in work associated with projects and an increased focus on essential elements of education that teachers felt related to student success. They concluded that many of the projects previously underway were perhaps less strategic than originally thought, and so they were canceled. This action (and the ease with which discussions of how to cancel or postpone projects were undertaken) is in contrast to the norms in many organizations, in both the public and private sectors, with new projects added to the existing portfolio until project fog settles in.

To cancel projects is often tantamount to declaring them a failure; but in a PPM-enabled context, that decision—to "reinvest resources in a higher strategic priority"—is properly seen as a positive and appropriate move, and cancellations are regarded as strategic rather than punitive decisions.

Another benefit of combining balanced strategic management and project portfolio management can be seen in the Peel example. As the leadership team was undertaking its annual planning process in 2005, it made the decision that if it was serious about impacting student success (as measured by standardized test scores), it needed to focus on finding out what schools with higher scores were doing that lower-scoring schools were not. This involved taking the analysis of cause and effect relationships to a whole new level. The result of this insight was a significant internal research effort that I supported to locate and then replicate "transformational practices." These were teaching or administrative practices in high-performing schools that were distinct and different—and which were validated as having an impact on student performance.

Subsequent efforts to publish and train other teachers throughout the system on these practices resulted in a significant increase in EQAO test scores system-wide in the following years. This effort, facilitated from the associate director's office and supported by a concise communications strategy from the Director, was credited with having a major impact across the system and is an excellent example of how allocating your resources to high-yield projects can impact strategy execution in a measurable way.

This example is testament to a significant culture change occurring and is a key finding. Over a period of about three years, the school board slowly moved from a traditional budget-driven organization that perpetuated the status quo to a more strategically oriented organization valuing actions that attacked the root cause of problems and challenges and resulted in a measurable effect on performance. This is a significant shift in any public sector organization, especially one as large as the PDSB. Much of this change came from becoming more focused on data-driven decision-making (simply by making sure the data about performance against target was generally made available to everyone in the organization, through the web). This is another notable learning from this case study—high visibility of results yields high organizational attention on actions designed to fix any gaps in performance.

STEP 4: STAYING THE COURSE

During the long process of integrating the Report Card for Student Success, there were times when it may have been tempting for key leaders to abandon this approach because it was taxing and difficult. But they did not. In fact, at one point mid-stream in a meeting with Judith, we discussed what the key ingredients were of organizational change—which installing PPM clearly requires in most organizations. You are not only changing the *what* (by clarifying strategy through measurement) but also changing the *how* (reducing projects, aligning effort and reducing the focus from the many to the few). This can be challenging. Yet Judith was particularly careful to continually communicate the benefits of this approach whenever she was out in the field; she often used slides like the one shown below to help her make this point succinctly and clearly.

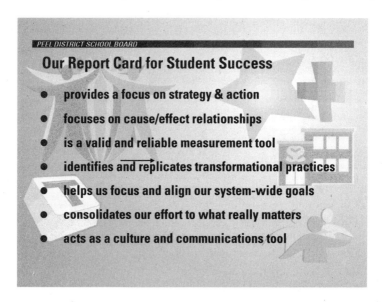

This helped the organization see that there was a clear leadership mandate to make this approach stick and that it wasn't simply a "flavor of the month." (Past efforts had been seen that way in the field, because of the lack of consistent support.)

So with a clear vision and effective peer relationships, a leader can eventually prevail over even the most prominent naysayer by remaining true and steadfast in the face of normal, natural organizational resistance. This requires a constant focus on two factors: employee communication and commitment. The execution of any organizational change benefits from an understanding of what to do when you communicate and how this impacts on commitment to the change. A simple picture on the following page tells the story about how to do this.

INTERESTING DATA POINTS FROM PARTICIPANTS

Throughout the case study, our team kept observations of the process and its resulting outcomes. We paid close attention to how decisions about the designs of various key processes were made and how successfully they were executed. This expert observation is a key part of how action research is conducted and evaluated. We also interviewed key leaders and conducted follow-up online surveys with participants about halfway through, to see how they were finding the process.

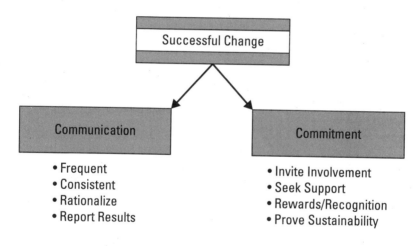

One of the things that we discovered: the longer a superintendent had served, the less onerous he or she found the time commitment required to master the new process. This same unsurprising correlation was seen between tenure in their jobs and the extent to which the RCSS process helped the participant understand the organization's strategy better. Less experienced professionals found the process of defining their organization's strategy in measurable terms more helpful than those with more experience. Yet, when I spoke with these experienced leaders informally during various sessions and meetings, they all seemed enthusiastic about the process and were supportive of its intent and outcome.

The surveys confirmed this. A critical contribution of the generic methodology is to provide a measurable clarification of organizational strategy and to enable the organization to select projects which align to delivering results connected to this strategy. To verify this, I collected "before" and "after" comparisons from participants about the clarity of their organization's strategy. In the case of the 19 respondents who answered these questions in Peel, there was an astounding 71% increase in the average clarity of understanding about the organization's strategy, among these leaders, after the first two years of the process.

Some comments (from both observation of the process and during the interviews with leaders) further validate this outcome:

- "The process initially seemed too difficult for us to agree on; but getting the measures right forced us to define what our focus really was going to be."
- "If we are going to pick better projects to fit our strategy, we first have to be able to measure the impact of any proposed project and then compare it to the targets and results we expect to have at a board-wide level."
- "The Report Card for Student Success will help me change my teachers' views of what's important and help them align their teaching activity to our goals."
- "This exercise has really helped me understand why our system feels so overloaded . . . we have too many projects with too many objectives and no way to sort out what we should really be doing every day."

In short, the findings of this first public sector case study support refining and implementing a strategic performance management system, right alongside your strategic project and portfolio management methodology. This yields a significant consolidation of effort and a resulting increase in performance.

CASE STUDY CANDIDATE 2

The subject of the second public sector case study is the Peel Lunch and After School Program (PLASP). As one of the largest operators of childcare centers and after-school programs in Canada, it is chartered to operate as a non-profit foundation rather than as a private enterprise. Its sources of funding include government grants for preschool children, income-geared tax subsidies from the province of Ontario, and the fees paid by some parents. It is another interesting example of a variant often found in the public sector worldwide, a non-government organization (NGO) or non-profit foundation that gets its sustainable income from both private and public sources. These types of organizations are often heavily regulated or influenced by public policy considerations but technically operate at arm's length from direct government control. From its marketing materials, the organization defines its mission as being "a world-class child care provider" with a focus on "child-centered development."

PLASP currently operates 17 nursery sites (providing care for infants from age six months to four years) and nearly 200 school-based programs that provide before- and after-school care for students in grades one to six. It has approximately 550 full and part-time employees and an annual operating budget of over $20,000,000, making it a significant enterprise even though its orientation is not-for-profit.

The CEO is appointed by an elected board of directors. The organization has a recognizable management structure, similar to those of private sector firms with a head office and field operations. It has an excellent reputation and brand recognition within the area it serves. (For instance, it has been consistently voted in media- and community-ranking surveys as an excellent parental resource and a top care provider in the region.)

WHAT WAS DONE DIFFERENTLY

In this case study, the organization elected for a broadly consultative process to increase the likelihood of broad employee buy-in to the PPM and Balanced Scorecard combined approach. In my experience, this is not uncommon in not-for-profit and NGO organizations, both large and small, where there is a heavy premium placed on consultation and staff consensus.

As a result, they assembled a cross-functional working team, including representatives from every major department/function in the head office of the organization and representatives from various school sites in the field. The resulting working group, while large at 22 members, was tasked with approving the methodology and approach on behalf of the entire organization, and developing the strategy map and defining the associated measures. They were to focus on the initial steps, up to the point where the draft scorecard, measures, and associated PPM processes were complete. They used the same theoretical framework described in the opening chapters of this book, and their approach to the work was almost identical to the previous case of the Peel District School Board (PDSB) where the only substantial variation made was in the language used to define concepts to suit their own organizational context.

In the case of PLASP, the team and I used a very simple two-phase approach that resonated with the group. It involved focusing on developing clear strategy statements and associated measures, then moving into the associated steps of defining projects to impact their goals, and then planning for action and implementation. This is best seen using the same phase-diagram that we used with the program participants:

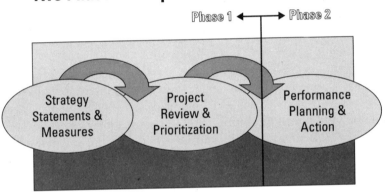

The focus of our work will be on:
• Defining our strategy and associating measures with strategic goals
• Completing a strategy map and defining projects and processes that matter
• Plannning for action to close gaps and over-achieve our goals in 3 years.

This initial work took the better part of two months to accomplish, primarily because of lapses in time between group sessions. (Meetings were tricky to schedule for the group.) However, at the end of this period, we achieved group consensus to proceed and began the work of creating a strategy map, balanced scorecard and its associated measures and metrics to help guide decision-making about projects, processes and other strategic actions.

Of course, as is often the case, this interim work was eventually "buffed and polished" by a graphic artist and laid out in a more engaging way for presentation outside the committee. For the purposes of this book though, it is more useful to see the actual form of the committee's work rather than the final version, created later, which you can see on their website at www.plasp.com.

In all group meetings, our team (three part-time consultants with additional expertise in process measurement, process design, technology and people practices) provided hands-on facilitation support for the group and acted as experts on the implementation of the generic framework, how to apply methodology, and creating some supporting tools for participants to use. We did not focus on content. This is how we normally work: it is the internal team's responsibility to be the experts on their own organization and its work; our job is to ensure this knowledge gets extracted, examined, discussed and documented in a useful form for use within the methodology.

Subsequently, the group re-assembled and in two sessions completed its remaining work. The result was the creation of something we now use with most of our clients, an "interpretation guide." This document highlighted each strand of the strategy map, clarified its definitions and goals, presented the associated strategic measures and sources of data, and outlined the initial targets. A sample page from PLASP's work is shown on the following page. You will see just how simple and clear the definition is (a self-assessment scoring tool), along with measures and targets (100% of sites assessed with a target of at least 90% achievement) and the baseline.

This same approach was taken for every single element of their strategy map; the organization did not have to rely on everyone interpreting the strategic goals and measures on their own. Rather, they wanted them interpreted the same way, consistently. I see this as a reinforcement of the value of consolidating effort and alignment of purpose through the selection of strategic projects that is so central to eliminating project fog.

PLASP's Balanced Scorecard Measures

~ Interpretation Guide ~

February – March 2006

INTERNAL PROCESS

PLASP's Scored Assessment Model (SAM):

Definition: Self-assessment checklist summarizing the best practice statements for each success factor (as derived from the Model Site definitions)

Baseline: None (new tool)

Target: 100% of Sites to achieve at least 90% on the Scored Assessment Model

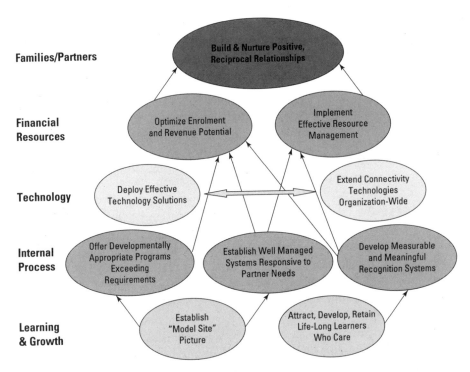

PLASP's Strategy Map
We Strive to be a World-Class Early Learning and Child Care Provider

We learned something else in this second case study. In Peel, the initial effort focused on the measures and targets, and this is important work. However, after a while it became clear that the lack of clear definitions of what and how we measure was creating a possible gap in understanding and interpretation which we later fixed. At PLASP, our team helped the client develop detailed definitions to accompany the measures and targets from the outset. This ensured that the committee's strategic intentions in selecting and defining the measures would be understood and respected across a very large and dispersed field organization operating in more than 200 individual sites. Here is another example (from their interpretation guide):

Major Special Events:

• **Definition:** A Special Event is a new or unique idea which is introduced as a variation in the program. Major Special Events are usually theme-based, involve planning with the children, and result in a series of activities that culminate in one large activity (e.g., Musical Production, Talent Show, Hawaiian Day, Circus Day). The objective of Major Special Events is to vary the type of activities offered in the program, to enhance the quality of the program and the children's enjoyment of the program, as well as to encourage the retention of children in the program.

• **Current Baseline:** The current expectation is 1 Major Special Event per term in each of the Before School, Lunch and After School programs.

• **Target:** 1 Major Special Event per month, in each program, which is communicated to parents through the posted activity planner, posters/notices in the program, and/or information flyers/invitations sent home to the parents.

• **Reporting:** Major Special Events are to be indicated on the monthly program activity planners sent into the office at the end of each month. The events are documented monthly, and a monthly report by school, by program, is produced. Parent Satisfaction for Major Special Events is rated by parents in the Annual Parent Satisfaction Survey.

Note the clarity and elegance of defining exactly what a "major event" is, how it is constructed and implemented, and how it will be measured. If you were a site manager in the field, this kind of specificity would be important if you wanted to ensure that your site met the organization's expectations. When we then connect these kinds of definitions to a self-assessment model (SAM), any site can determine how close it is to a "model site"—that is, one operating at 100% of the standards of excellence in all areas.

By the way, we helped PLASP develop its "model site" based on the learning we got from researching and implementing transformational practices in Peel. The model site exercise was a powerful internal search for better practices that could be replicated throughout the organization to improve results. If a particular site was doing exceptionally well on a measure, the team visited the site, discovered what was being done, validated that it made sense and could be replicated and then documented the approach for others to benefit from. This created positive internal approval for those creative and innovative site managers who were exceptional performers already and also helped those who were facing performance challenges to realize that it was possible to find new ways to change current baseline performance levels. This helps empower employees to make a difference and feel good about aligning to the purpose.

CONNECTING INDIVIDUAL AND CORPORATE PERFORMANCE MANAGEMENT

As we pointed out in the opening chapters of the book, PPM cannot be successful unless you connect it to other core processes in your business. One of the most obvious examples of this is the annual or ongoing employee appraisal or evaluation system. Most organizations have some form of this kind of individual performance management system. It is often connected to compensation, and perhaps to talent management or succession planning efforts as well. However, what I often find when I consider these systems in a client context, is that a massive amount of effort is potentially wasted unless these are tied to measurable performance indicators.

All too often, the employee performance evaluations are connected to vague competencies, long lists of accomplishments, or to some read on their likability or fit within the culture. While some or all of these may be important factors to consider, what is most essential is that organizations learn to connect *organizational performance management* to *individual performance management.* Ultimately it is the cascading of clear and simple tasks that will support strategy execution that helps an organization move toward its goals. Therefore, once PPM is established and those strategic goals are known, they must be directly connected to what employees are rewarded and recognized for. Otherwise, there is a potential disconnect between what employees are being told is strategic at an enterprise-wide level and what they do day-to-day.

To avoid this at PLASP, the midstream identification of this as an issue caused the vice president of human resources to reach out to us and find a way to make annual employee appraisals directly align with the strategy measures and goals. The result of this effort was a new system of employee appraisal that was met with much enthusiasm by both leaders administering the system and employees being appraised. See an example of a section of this new employee performance management system on the following page.

INTERESTING DATA POINTS FROM PARTICIPANTS

Information about this case study was collected in a similar fashion to others we have described, including the use of simple pre- and post-surveys about how participants responded to the approach, interviews and notes taken during key meetings or events and tracking of whether there were measurable improvements in performance levels within the organization. This is ultimately the key concern of most CEOs and executive teams, and indirectly of project managers who support them with effort designed to improve organizational results. So let's see what they had to say.

Working with your Program Director or Program Coordinator, write 2-3 measurable goals for yourself that will support the achievement of Balanced Scorecard Goals in any of the following measurable areas. Actual Results will be completed and rated at the end of the next performance cycle.

Scorecard Measure	Objectives for Coming Year	Actual Results Achieved
% of Annual Parent Satisfaction		
% of Annual Child Satistaction – if applicable to your program		
Average Program Vacancy Lag Time		
% of Enrollment to School Population		
Program site achieves 90% on PLASP Scorecard Self-Assessment Model		
Major Special Events held monthly		
Mini Special Events held weekly		
# of WSIB reportable staff accidents	Perform all work in a safe manner on a daily basis to ensure that I am accident free.	
Other Objectives not directly related to the Balanced Scorecard		

We interviewed three key executives (CEO, CFO and VP, HR) as well as the assigned internal project lead (director of field operations) at the end of the process, and all were enthusiastic about the outcome of the multi-year effort. The CEO said, "Every time I see something going on now, I ask myself how it can be measured or improved . . . and I know everyone in the organization does too. This has taken our organization to a new level of maturity in terms of managing our own performance." The results of the interviews provided lots of anecdotal, supporting or clarifying commentary about how the process had generated significant insights for leaders around how to connect effort, performance and results. So this continues to be a key benefit of the PPM process.

The survey results make this benefit even clearer where the group felt strongly about the contribution and value of the PPM framework we had provided to them. Here is a summary of some of the more important findings:

- Of the respondents, 65% strongly agreed with the statement "There was an organizational benefit to me participating on this team."
- More than 50% of respondents strongly agreed with the statement "We would recommend this process to another organization as having high value."
- Similar high ranges of agreement can be found for statements such as "I think the process is sound," "I think the process applies to us," and "I think the process will generate results for us" (73% either strongly agreed or agreed with this last statement, for instance).

The only area of concern that emerged from this case study is the trade-off question between process validity and complexity. This was determined as important because only 26% of respondents strongly agreed with the statement "The return from the BSC exceeds the effort spent." This reinforces earlier discussions in the opening chapters and other case studies about balancing complexity and completeness versus realization of project benefits at a reasonable investment level of time and effort. To address this, I often stress with clients that any balanced performance measurement system is never going to be perfect, and so the objective should actually to get it to "good enough." (Or, in my role as professor, what I often say to my mark-hungry students is that a solid B+ will do it every time!)

One research finding was of particular note for me in this case study. Among all respondents, 40% felt that the "most impact" from the intervention was "the ability to state and measure strategy clearly." This finding correlates directly to the underlying hypothesis of this book and is central to our conclusions that designing and implementing effective PPM process in any organization drives clarity about the strategy down to the lowest levels of an organization. This is a critical and well-researched factor in the successful execution of strategy; CEOs in particular should take note of this. In fact, in analyzing the PLASP survey data, the following conclusions are easily drawn:

- There was a clear improvement of the understanding of PLASP's strategy in participants before and after their participation in the BSC and PPM process.
- Participants had a feeling that PLAPS's business processes were *efficient* before the intervention but had an even stronger sense that they would improve post-implementation in regard to both *efficiency and effectiveness.*
- Participants had an absolute sense that PLASP had too many projects under way at the same time prior to the intervention and a sense that this challenge was addressed by the methodology (a finding replicated in the previous case study as well, and which generally is seen as a benefit of applying the methodology).
- And there was a sense among participants that after completing their strategic measures, they would have an improved ability to state which projects were more or less strategic to help their organization execute its strategy.

In addition to all of these findings from surveys and interviews, here is what is ultimately the most important thing to consider in assessing the value of PPM to your organization. Absolute performance has increased and, relative to their targets, PLASP is at or above their targeted level of performance at the three-year mark. While some indicators still have not been achieved, the focused and aligned effort of the leadership team and every single employee within this organization being motivated to over-achieve is bound to get them there. And that matters.

CONTINUOUS IMPROVEMENTS

No organizational change is ever complete to the point where an organization can afford to remain static. As an organization evolves, it must re-invent itself and continually challenge itself to improve its quality of execution to remain globally competitive. That is not any different in the public sector—there is always the challenge to do more with less and find creative solutions to tough problems.

With PLASP, we checked in just before the publication of this book to see how they were doing. Of course, one thing I wanted to see was if they had stayed the course and were still using the methodology. To my delight, I discovered that they were and that, in fact, they had continued to push the boundaries of performance within their organization by doing a couple of new things that were having a significant impact on employee performance.

One thing that we had urged PLASP to do as we wrapped up our initial work with them was to find a way to get performance measurement information out to the field in real time. This was a significant challenge, given the scope, scale and nature of their business. However, what they did was create a very powerful website that any staff member could access any time it was convenient, that reported real-time performance results.

This simple tool, graphically easy to understand and not hard to navigate, was created internally by a small team in a few months. It is an excellent example of using simple, automated tools to enable the PPM process. A snapshot of the home page and an example of how information is displayed underneath when you click through is shown below. It is interesting to note that both Peel and PLASP report that simple access to accurate and timely data improved both performance and quality of decision-making in regard to alignment of effort to strategic outcomes and defining and selecting strategic projects.

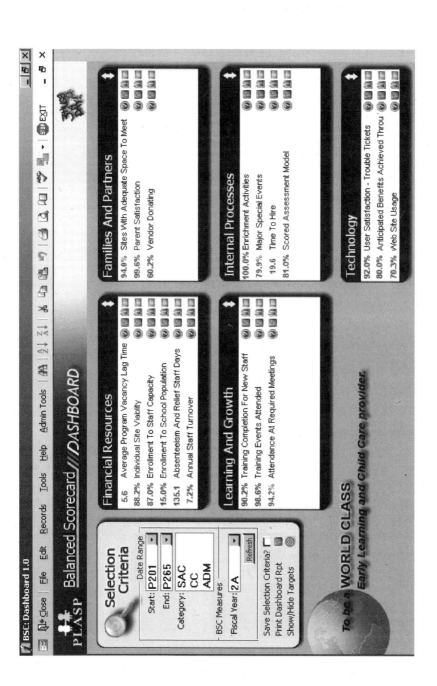

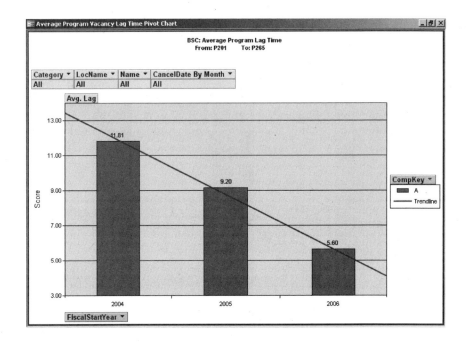

WHY PPM REALLY WORKS

In summarizing the findings from the two public sector case studies, it is clear that the generic PPM methodology creates clear and compelling benefits for those who use it. But why is this so?

One hypothesis, not new to many of you or to others who have written about this phenomenon, is that the management priority—effective and productive leaders who perform well above expectations—is quite different from having people who simply perform at expectations. This is to be expected: those who act differently get a different result. A picture of this finding will be more useful to a reader than a lengthy description:

The Productivity Hierarchy

*To do "more with less", an organization and its leaders must focus its energy on highly leveraged activities. This priority of **process over people or task management** ensures higher productivity and more efficient use of organizational resources in any setting.*

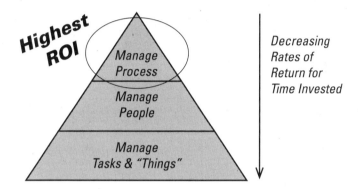

This concept can easily be found in both of these public sector case studies. In each instance, the leadership decided that a new way of doing things was required, and they opted to work at the process level to change results rather than focus on tasks, things or their people. Most people want to do a good job and, properly trained, can do most jobs assigned to them. What makes a focus on process so important is that it finally helps us define that the work assigned is the most relevant and important contribution to strategy that we can think of. That is what PPM is all about and this is the real contribution of effective C-suite executives and great project managers alike—the consolidation of effort and alignment of purpose!

## CHAPTER TEN

Where to Go from Here?

Hopefully you have found this book stimulating as you think about the connection between strategy, project selection, and organizational performance. While certainly more complex than most practitioners and executives initially believe it will be, a workable PPM process framework is something you can define, and it will make a substantial difference now that you understand the theoretical and practical underpinnings of PPM.

To bring this full circle, let's return to the three benefits of an effective PPM process:

1. Improves the ways projects are conceived and submitted for early approval to avoid wasting effort on non-strategic projects.
2. Defines practical methods to directly link project outcomes to a balanced set of measures that optimizes and aligns the project portfolio with your strategy.
3. Defines how to govern and manage the project portfolio with re- spect to clear project prioritization, staged activation, tracking, and reporting to realize maximum benefit, control costs, and reduce risk.

If you sequentially followed the content in Chapters 2, 3 and 4, then you can see how the generic process framework works. By adapting this to your own organizational context, you can use it to create similar benefits for your own organization. My objective in creating a generic process map for PPM that could be universally applied was both to present a theoretically sound process framework and to enable it with specific methods and tools that work in practice. And, as we've seen, these benefits can be accomplished.

Regardless of your particular role within an organization (board member, executive, project management professional, consultant or vendor/partner) you now know how you can help your organization make more effective project selection decisions. This is a potentially valuable contribution. Yet knowing how to do it is one thing, but doing is another. As we discussed in Chapter 7, it takes courage to act on these insights in order to produce new results.

Awareness of the need to change is the first step on your way to successfully implementing change. This book should have stirred your awareness of the major issues in terms of creating, selecting and executing strategic projects that matter, and of what an organization needs to do to accomplish these tasks while balancing effectiveness and efficiency.

Perhaps this book has helped you see the need for stronger executive sponsorship or more appropriate board oversight of high-risk projects; perhaps it has created a need to clarify the types of strategic measures you use to guide the way. Or maybe you've become aware of systemic or organizational design issues—the role of your PMO; the degree of centralized control—that are potential barriers to the successful implementation of PPM within your organization. No matter what you have identified that needs to be changed, it is my hope that this book will get you going.

NEXT STEPS

To help get you started, the author team put together a list of three vital questions we would use if we were in your shoes to evaluate your next steps. They are a starting point:

1. Can you find one or more sponsors (including yourself) with an interest in implementing a strategic PPM process within your organization?
2. What would your organization gain from using PPM, and what can be done to realize these benefits quicker?
3. Is your organization willing to be truly strategic, to define measures associated with their strategy, and then to demand organizational accountability for execution of the plan?

If you can identify positive answers to these guiding questions, perhaps you are poised to take on the leadership of this change within your organization and make a real difference. If you are an executive, perhaps you will be able to creatively engage your project management professionals to help you gain traction. If you are a board member, what aspects of strategic measurement can you help sponsor to make PPM an easier thing to accomplish within the organization?

As we have seen, PPM can become a defining competitive advantage for any organization. It works in the private, public and not-for-profit sectors. Every organization can benefit from the proper implementation of the PPM process. (Some organizations seem unwilling to venture into this uncharted territory. If you belong to one of these organizations, it may be more challenging to lead a change such as this. For those of you caught in this situation, I counsel continued courage and persistence.) Or perhaps there is a less-scary starting point. Perhaps you can add an on-strategy measure to your project submission template. Some progress is better than none. So keep at it. Something will eventually take hold, root and grow to make your contribution successful and visible.

Also keep in mind, the journey to implementing an appropriate PPM process takes time: time to socialize the concepts among executives and project managers in the organization; time to design the specifics of your own most appropriate strategic measures associated with project scoring tools; time to implement the process, retrospectively and going forward, in order to gain confidence in its reliability and performance. Implementation is often measured in years, not months. The process is not easy, but it is powerful, so, like any process of fundamental change, expect it to take longer than you think and plan for!

Executives in your organization need to understand what they are signing up for when they approve your initial proposal. While they can theoretically understand what they are agreeing to, at this stage they may not fully grasp that a truly strategic project selection process puts to rest many of the ineffective, but familiar, methods of project selection used today. Factors like executive intuition, inter-departmental rivalries, politics and personal power will be diminished as they progress down this path. While executives must be seen as supporting strategic initiatives, you may be surprised by the level of resistance of some executives when they begin to realize just how much of what they do will change. Even scarier for some may be the notion of being accountable for measurable, strategic results, especially if this has historically not been the case within the organization. If you are a project manager trying to get PPM implemented, remain empathetic to your executive colleagues. Like all of us, they need time to adjust to change, so be patient.

Finally, if you are a project management professional, we urge you to spread the word about PPM. Make sure that when you attend functions, meetings or conferences you explain to those who are misusing the term "project portfolio management" what it really means. Challenge those who are making it too simple by distinguishing between enterprise or program management versus the truly strategic processes of dynamic project selection and portfolio management. Encourage those in positions of influence that, while potentially complex, the rewards of doing this right and making project managers strategic partners can really make a difference to strategy execution. And when presented with the opportunity to be strategic, embrace it and run with it, because it's a professional gift.

I hope this book ends on a high note for you—the anticipation of being able to make your work matter. You now have the tools to begin your own journey into the PPM process in your workplace. Any of us would love to hear about your progress and answer any questions you may have about PPM. Please contact us at www.projectgurus.org.

Review of Relevant Research and Annotated Bibliography

Among students, academics and other researchers reading this book, questions may arise about the study of project fog and what has already been concluded and written about it. And I am often asked by practitioners how they should consume and apply research in our field. They wonder: is the exercise of understanding academic findings even useful in practice?

Now that I have encountered these questions so frequently, it has become my mission to ensure what I do matters to those in positions of leadership in the professional practice of project management. I recognize this is a lofty goal, but unless ideas have an impact on practice, research is a wasted effort. The starting point for achieving this outcome requires research that is rigorous and reliable. If this can be accomplished, there is potential for the research findings to be relevant to and usable by practitioners. But being rigorous alone is no guarantee of this outcome.

The debate about whether research can be both rigorous and relevant is longstanding in the academic community and particularly in the IS/IT field, where it has been debated intentionally in recent years. The emerging conclusion is that it is possible to accomplish both of these objectives simultaneously so long as the product of the research provides implementation suggestions and acts as a stimulus for action. Why is this last point so important?

Lyytinen (1999) refined the concept of relevancy in systems research by suggesting that research needs not only to be accessible to practitioners but must also produce a long-term change in behavior. This is particularly important in a project management setting, where individual behavior has such a substantial impact on project outcomes and the execution of organization strategy. It is the collective work of all project managers globally that in turn defines the state of the project management profession. So in conducting my research, I intentionally sought to gain insights that can affect practice globally.

Once awareness exists, the stage is set for the professional to question current results and be shown new ways to act (for instance, by implementing a new system, approach, methodology) targeted to address gaps in their current practice. It is the stimulation of an awareness of a gap in practice that causes a yearning for learning!

To help you on your journey, I have summarized some of the most important research in our field in the following pages and included a bibliography that will help you locate those of interest to you, organized by major topic.

STRATEGY REFERENCES

Of primary importance to this work are questions of strategy in an organization. The clarity of the strategy is important; so is its ability to influence the actors in an organization to act accordingly, and the validity of the processes used to ultimately decide upon and document the strategy in both a private and public sector context. These types of questions have long been studied by range of notable authors in this field, such as Steiner, Porter, Prahalad, Hamer, Mintzberg, Rumelt and Teece, and many others. Their emphasis is normally on issues related to strategy formulation in a business context, many including commentary on the specifics of organizational strategy and strategic nuance in the public and not-for-profit sectors as well. Having surveyed this extensive body of literature, I believe there is ample evidence in the literature on what constitutes effective strategy formulation and an equal number of suggestions on techniques and tools to use for undertaking any kind of planning exercise. It would be redundant to recite the full range of literature on this broad topic here except to say that it should be clear to any reader that having a well defined strategy that is actionable is an important component of success in any organization.

In fact, the essential element of strategy of interest to this particular book is not its formulation but rather its execution. As noted previously, selecting and executing projects that are truly strategic rather than simply financially efficient is not as clearly understood as it should be (Morris & Jamieson, 2004). As we contemplate doing so in the context of very large organizations (in either the private or public sectors) with huge numbers of possible projects to choose from and highly complex strategies to manage, the problem does not appear to have been specifically studied much at all. So it is important to look beyond the large volume of general writing on the formulation of strategy in order to gain insights into the more complex, but less studied, task of executing strategy in an organizational context using projects as the building blocks of strategic execution.

An equally important area of specific interest to us is contained within the domain of business strategy, especially on focused academic literature on IS/IT strategic alignment to business, and particularly in a project context. Obviously, since questions of managing projects often originate or reside in IT in many firms, innovations in terms of managing projects strategically may emanate from this source of knowledge. One of the major contributions in thinking about Business-IT alignment was presented by Luftman and Brier (1999). Their *California Management Review* article in autumn of that year considered survey data from over 500 firms in 15 industries and attempted to define the six enablers and six inhibitors in an exemplar organization. Their conclusions are often referred to in subsequent articles on this topic.

It is not surprising that the early literature emphasizes themes such as management acumen, leadership support and project prioritization as being critical to successful business-IT alignment. Indeed, in their seminal work, Henderson & Venkatraman (1993) emphasize the need for IT to evolve from administrative computing "toward a more strategic role that supports the organization of tomorrow" (p. 4) and provide a suggested framework to accomplish this. In more recent work (Ciborra et al., 2000) the theme of globalization of the IT infrastructure in relation to the needs of ever-larger global organizations has emerged as an important consideration for those studying business-IT alignment.

These are enduring themes that are also repeated elsewhere (Luftman, 1996; Earl, 1993; Chan & Huff, 1993; Liebs, 1992; Wang, 1997). And very clearly, effective strategic outcomes at the firm level always rely heavily on questions of leadership and management effectiveness (as extensively explored by others, including Zalenick, 1977; Bennis 1989; Briner et al., 1996; Yukl, 1998; Kotter, 1999). Many of the authors cited above note that an absence of leadership almost certainly translates into a sub-optimally effective organization in general and not just in project management terms.

In their groundbreaking work, Morris and Jamieson (2004) offer leadership core competencies based on an aerospace case study (figure 2.7, p. 30) that summarizes distinctions between the leadership competencies of more senior (Project Director) and less senior (Project Manager) employees. It certainly confirms a conclusion that project management success is more about leadership than it is about management.

So, while it is quite easy to locate sources of additional information on the literature in the area of leadership and management acumen in relation to project management outcomes, the area of strategic project prioritization and selection of strategic projects is much less explored and does not seem to have attracted the same level of attention in the literature. The question remains: how do we get the strategy accomplished once it has been formulated? This is often referred to as strategy execution and is distinct from strategy formulation. In project terms, it could be expressed by the questions "How do we pick projects to support our new strategy?" or "How do we align PM resources to focus on critical strategic outcomes?"

Obviously, prior to being able to commit to "strategic" project management practices (including potentially PPM), it would be important to be able to measure the strategic contribution of a single project, and important to understand strategy at the enterprise/organizational level in precise measurable terms. In the existing literature, we see parts of this as an early focus on "success criteria" or "benefits realization" by linking project success to externally referenced deliverables (Turner & Cochrane, 1993; Waterridge, 1998; Wells, 1998). But more recently, other authors have noted the dearth of insightful studies that can adequately define "strategic project management" (Artto, Martinusio & Alto 2001; Benko & McFarlan, 2003; Morris & Jamieson, 2004; Morris & Pinto, 2005) and what must be done to convince senior executives of the value of making investments in improved project management practices. Morris and Jamieson (2005) present several case studies from a variety of industries and point out that the maturity of strategic project management practices varies by industry and may be perceived quite differently (i.e., construction versus pharma-biotech). This must be taken into account when considering the application of advanced project management practices like PPM in different sectors of the economy.

Intuitively, it would seem obvious that for projects with longer time horizons, the more likely cause of a perceived lack of strategic contribution at project completion may be rapidly changing industry or business circumstances. To address this issue, we must be able to tie project outcomes to strategic goals and stakeholder expectations (Tuman, 1986; Davis, 1995; Belout, 1998; Baccarini, 1999). This causes anxiety among managerial decision-makers because they would likely not agree on what is actually "strategic" and whether or not a project was on track to deliver its strategic outcomes at any point in time. This is often because strategy is not expressed in measurable terms (Mintzberg, 1994; Norton & Kaplan, 1995). This problem most likely occurs in "mega-project settings," for instance, the implementation of Enterprise Resource Planning (ERP) or Customer Relationship Management (CRM) technology, where the time horizon for such a significant IT-related project may run into years and touch all core operations of a company. In these types of projects, it is critical that the project be directly connected and remain connected to the organization's emerging and changing business strategy and that managerial decision makers have the ability to examine this and draw joint conclusions dynamically and continually.

Previously, and in response to this particular challenge, I explored the connection between using a strategic measurement system such as the Balanced Scorecard in a single "mega-project" setting to help maintain the connection between project results and strategy over sustained project durations in an article for the PMI research journal (Norrie & Walker, 2004). This article builds on the profession's continued evolution, which is moving beyond a mere cost-time-quality view of project management work to a more strategic view of project management—such as the one proposed in the Logical Framework Method (LFM) (Baccarini, 1999) and others. While LFM may be seen by practitioners as useful, it does not address the next task required for the field's progress—extending this thinking to multiple-project or enterprise settings at the broader organization level.

If we want to extend this approach to the enterprise level, it would suggest that any effective methodology for the implementation of any Project Portfolio Management system would need to consider the nature of dynamic and constantly changing business environments. This can involve team members who do not necessarily agree on the interpretation of strategy in relation to their assigned goals and objectives. This problem has been explored in the past, most notably by Bennis and Nanus (1997), Bennis, Spevietzer, and Cummings (2001), and Turner and Cochrane (1993). The existing literature contains both general team-based solutions (Katzenbach & Smith, 1993; Robbins & Finlay, 1997; Yukl, 1998) and specific project-based solutions (Briner et al., 1996; Thite, 1999). By examining this problem in some detail, it is evident that the connection between project outcomes and strategy is either ambiguous or understood by only a few key stakeholders, rather than more broadly accepted by everyone who influences the full range of project outcomes. Again, this suggests the presence of decision-making and communication challenges around the entire notion of organizational strategy. Many researchers, particularly Senge (1990), have stressed in their work that a narrowly held vision is insufficient in most leadership contexts and fails to create purposeful co-ordinated action among all followers.

To further allow for strategic trade-off decisions that might include choices such as delaying or canceling existing projects in favor of newer initiatives that have higher strategic value for the same expenditure of resource, it is essential to develop a method to effectively express organizational strategy in measurable terms. This is often referred to an "evergreen" approach to strategy. To enable this outcome, executives would need a method that permits agreement on these decisions and would be more robust than simple "gut feeling," otherwise, the process could descend into political chaos as each executive fights for his or her own particular point of view. In fact, it is clear that who proposes projects is often a factor in their selection in organizations with an absence of any strategic project selection system (as pointed out in the book's earlier chapters).

But this is not an appropriate decision-making method. Nonetheless, it is common that the only manifestation of organization strategy that is seen to be reliable is found in the actions made by the executive team (Kotter, 1990). This limits the ability of the rest of the company's employees to make appropriate strategic decisions without direct executive input. Therefore, any strategic measures or models that enabled a more highly articulated level of understanding about strategic intent among the organization's employees would be of general benefit to the firm in executing its strategy. While on occasion it may be partially contemplated in some existing implementations of PPM, it is not evident that the objectives of clear communication among decision-makers, coupled with a strong connection to underlying strategy have been accomplished in any systematic way within a described methodology to date (Morris & Jamieson, 2004). Given what is at stake in most cases, it is not prudent to leave this to guesswork; thus the purpose of this work is to explore and resolve this dilemma by proposing a possible solution to the issue of strategically oriented project scoring, selection and prioritization.

PROJECT MANAGEMENT REFERENCES

The literature on project management is extensive and is growing at a rate proportionate to interest in the field among both practitioners and academics. The literature often focuses on the operational or individual task level, where authors try to distinguish between individual project management techniques and multiple project management techniques at the enterprise level. However, there is also little doubt that most executives see project management as a tactical, rather than a strategic, contribution to their organizations (Thomas & Jugdev, 2002) thus giving rise to the problem outlined in this book in the first place.

All of the general literature related to the effective management of projects on time, cost and quality (often referred to as the triple constraint or iron triangle by practitioners and addressed in Chapter 2) uses methods to determine critical success factors in project management (e.g., Martin, 1982; Zells, 1991; Drummond, 1998; Baccarini, 1999; Byers & Blume, 1994; Clarke, 1999; Forsberg & Mooz, 1996; Whitten, 1995; Wateridge, 1999; Shank, Boynton & Zmud, 1985; Cooke-Davies, 2002; and many others). Articles or case studies that deal with single or multiple projects in a specific industry, while still important to the profession, are of less relevance to this particular work than those that deal with a holistic view of general project management processes. Of particular interest are citations that include early stages related to existing project selection techniques—the problem of study in this book. This is implicit in the terms of reference of this work, where the concern is on selecting strategic projects, not on managing them. Based on recent efforts within PMI to develop and publish OPM3 (their own project management maturity model), if an organization can locate a methodology to pick the right projects, there are ways of ensuring sufficient capability and maturity in the firm to actually execute the project itself. Therefore, the remaining concern would be to ensure the assigned practitioners are already experienced in effective single and multiple project management techniques. However, there is scant research on this particular focus area (project selection) and others (Belassi & Tukel, 1996; Archer & Ghasemzadeh, 1999) have already identified the struggle we will have as researchers if we are to engage in a conversation about "strategic project management" when the difficult issue arises of establishing the specific strategic value of a single project. Dinsmore (1998) clearly identified projects as strategic building blocks and notes the importance of selecting them appropriately, but he did not specifically define an approach to scoring projects that would be useful in a PPM process for the public sector and only referenced economic criteria for the private sector. So, this book appears to be at the apex of this problem at a point in time when the profession is struggling deeply with these issues.

Another critical area of the body of knowledge in project management related to this work is benefits realization—or, more colloquially, how we determines if a project has been "successful" using post-project reviews and comparisons to the originally proposed business case. While there are varying views on this in the literature (Busby, 1999; Baccarini, 1999), if one is to make relative comparisons between individual projects to try and make project selections, a determinant of what constitutes the expected benefits and outcomes of the project (both financial and otherwise) in clearly succinct and consistent terms would have to be present in each project proposal to enable trade-off decision-making. Researchers are only beginning to deal with this important question (Turner & Cochrane, 1993; Cooke-Davies, 2002; Jugdev & Thomas, 2002; Walker & Nogeste, 2004) as they are starting to suggest models for categorizing and comparing benefits realization between projects using a consistent framework.

One important element in any major project is its originating charter—a tool used to improve the clarity of project outcomes (Lavence, 1996). Similarly, others have concluded that a good project charter also improves communications about a project (Hartmann, 2000; Gioia, 1996). The most recent studies to emerge suggest a possible correlation between a poor project charter and increased risk of project failure (Christensen & Walker, 2004). This confirms the importance of implementing a clear and usable project charter in the planning stages of any project, as it can serve to substantially guide project execution.

This conclusion may also be related to earlier observations of attempts to tie project success to business strategy in ways that are visible to project team members. I agree that project failure can be the result of a poor charter; but it may also be possible that a weak articulation of business strategy in the first instance challenges any effort by project managers or executives to create clear project charters. This further amplifies the problem of aligning projects to strategic measures that is at the core of this book. So it would seem that identifying the charter process as the singular problem risks an oversimplification and therefore suggests that bad project charters are a symptom, rather than the cause, of the underlying problem of strategic ambiguity.

While there is some tangential literature related to the activation of projects to reduce capacity and resource conflicts, it is associated with creating and managing a Project Management Office (such as Crawford 2002), rather than focused on PPM. But it is important to note that of the literature used for this work, this interest seems to be quite emergent and relatively new (i.e., within the last five years). So while this supports the emerging interest in the field around these topics, it does not illuminate practice to any great extent. Nonetheless, the inherent assumption of note is that the selection of the project portfolio is assumed to have already been done, so the role of the PMO is then to ensure that projects are executed properly at a tactical level.

"Ensuring that projects get done correctly—the real goal of the PMO—is a different matter than ensuring that the correct projects get done" (p. 47) is a quote that appears in the February, 2001 edition of the *PM Network* (the publication for members of PMI) in an article titled "Choosing the Right PMO Set-Up." Similar issues around the proper construction of, and the variety of possible configurations of, a PMO in relation to project performance have been identified in the literature previously (Hobbs & Aubry, 2006; Dai & Wells, 2004). Once again, we see the profession's apparent emphasis on execution rather than the strategic selection of projects. This article, written by William Casey and Wendi Peck, describes a now common set of labels for various types of PMOs (using analogies to weather stations, control towers, etc.). The entire issue that month was devoted to the "emerging question" of what a PMO is or is not and how it should be set up for maximum strategic advantage. This lends credence to the relative newness of these concepts within the profession and the challenges project managers face when trying to ascertain how to move to a new level of strategic contribution.

Nonetheless, firms that have moved extensively into project management as an organizational modality often turn to creating a PMO as their next evolutionary step (Redmond, 1991; Raz, 1993; Butterfield, 1994; King & Anderson, 1996; Munns & Bjeirmi, 1996). Few of the cited sources identify the issue of strategic measurement specifically or delve into the issue of project prioritization or portfolio management in a meaningful way, other than to suggest that the need to prioritize projects exists (and the implicit assumption, once again, is that the firm knows how to do this). As previously stated, strategists have surmised for some time that prioritizing is critical (Luftman, 1999); so the conclusion that we need to do this is obvious—rather the question remains for practitioners, How to do it most effectively?

As far back as the mid-1980s, Tuman (1986) and Cleland (1986) concurrently recognized—and simultaneously presented findings—that contradicted the then-common notion that on-time, on-budget, and on-quality were the most strategically important and valid measures of project success (the traditional triple constraint so often imbued in project management practice). Yet almost two decades later, professionals and academics alike remain quite committed to these concepts and the pervasiveness of the triple-constrained model of project management is evident in the literature. This dependency may indicate the lack of a definitive alternative to this traditional model. At the project prioritization level, this means that we often use techniques that influence us to choose projects we think might be "successful" on this basis.

Other researchers (Turner & Cochrane, 1993) propose matrices and methods to deal with projects where goals are less clear and where traditional project management methods might fail to deliver as a result. We could surmise that projects perceived to be at high risk for failure might not be selected to proceed, even if those same projects were perhaps strategically essential (Dinsmore, 1998). The literature seemingly assumes that firms know how to select and prioritize projects appropriately.

However, there is some emerging research (Schwalbe, 2001; Norrie & Walker, 2004) challenging current thinking on the triple constraint and instead proposing the notion of "on-strategy" as more relevant to project management success. The concept is noted as "emerging" because this novel concept has not yet made it into the mainstream project management literature or the more recent textbooks (e.g., Kerzner, 2006), all of which fail to mention this when referring to the traditional constraints of project management—with the exception of Schwalbe's new text.

Learning to distinguish between influence and control over project management decision-making often means the difference between temporarily controlling an outcome by forced compliance versus actually creating a lasting change in people's behavior (Greiner & Schein, 1988; Kotter, 1999; Loosemore, 1999; Pinto, 1998). The literature suggests that project managers eventually come to the conclusion that they cannot oversee every decision to ensure that their project team members conduct themselves appropriately in performing their roles and realizing the project. So most project managers revert to some kind of exception-based or situational leadership method to address ongoing challenges, as recommended by established theory (Hersey, Blanchard, & Johnson, 1996). While somewhat effective, this tactic does not completely address the issue, because the traditional triple constraint deals only with decision making related to on-time, on-cost or on-quality project delivery. This continues to leave a gap regarding how on-strategy decisions are made within a project team and the role of the project manager in making this happen.

On this point, we turn to the literature around trust and its link to commitment and effective decision making (Meyer & Allen, 1991; Lewicki, McAllister and Bies, 1998). This topic may be addressed uniquely in reference to project teams but, in fact, it shows up in most organizational settings (including both the private and public sector contexts). Normally, there is a positive correlation between enhanced trust between employees and employers (often represented by the proxy of the project leader/ executive sponsor and the project team), as noted in the Lewicki et al. (1998) study.

Of equal importance is the conclusion that when trust is high and commitment is high, there appears to be improved decision making because of an openness to expressing, exploring and adopting alternate points of view in any particular discussion or debate ultimately leading to a decision (Bennis, 1989; Bennis & Nanus, 1997).

One of the factors that impugns trust is an over-reliance on hierarchy and positional authority as defining who makes a decision rather than on the process of how an optimal decision should be made. The emphasis in the literature on the priority of process over structural authority (Bartlett & Goshal, 1995; Pinto, 1998) is understandable; this approach generates improved outcomes and ultimately, the successful execution of a project. From direct observation of more than 100 project managers and participants of varying seniority during this study, it can be seen that the most competent project managers build strong, trust-based relationships with project team members. They listen attentively to recommendations and challenges from team members at the point when they are required to make a decision. This does not mean divesting complete authority or ignoring structure; good leaders accept ultimate responsibility but also enable and empower their teams to help generate a collectively successful outcome, for which they are prepared to share the credit. Yet, they accept personal liability for failure if something goes wrong—an updated version of the "captain going down with the ship."

Task and process management at the simple project level must be rendered relatively easy to codify, learn and apply if one is to consider project management a profession. The Project Management Institute, the largest certifying body for project management professionals in the world, publishes the currently defined best practices in the discipline in its *Project Management Body of Knowledge* (the most current version having just been published), which clearly focuses on the need to apply and measure adherence to standardized steps in the management of a project. The notion of the importance of project-based measures (and associated tracking and reporting of the work associated with project management) has been noted before (Hartmann & Jearges, 1996; Kiernan, 1995; Thamhain, 1994). Again, while this basic principle is valid, there still seems to be a need to ensure that the measures being used to track project success are truly strategic and not purely operational. Otherwise, the risk of delivering projects on-time and on-budget but with limited value may arise.

However, appropriate leadership decision-making at the strategic level is altogether different from task or process decision-making, even with the inclusion of strategic measures as a guide. In their landmark work, Project Leadership (Briner et al., 1996, p. 67) the authors emphasize the role of a "sustainer" as a key aspect of successful project sponsorship. They also stress the need for project managers to orient themselves towards alignment and away from enforcement—an elusive concept that entails the creation congruence among the team and with the project's goals by using a variety of activities and sources of power to influence others to act in accordance with the project leaders' desired outcomes, rather than relying on a traditional command-and-control management orientation.

Similarly, a colleague just completing the DPM program is exploring decision-making models in project settings and talks about the "paradox of control" arising from an effort to over-control project execution at too micro a level. Rather, the use of strategic measurement and strong leadership practices can encourage proxy decision-making from each team member in their individual sphere of control. This helps align decision-making to the overall project goals and improves project execution. But this can only occur if the individual project team member has a thorough understanding of the project's strategy and its connection to organizational goals and objectives (Bourne & Walker, 2005).

LEADERSHIP AND CHANGE MANAGEMENT REFERENCES

A review of the current literature in this area reveals that numerous projects are perceived as failures due to poor leadership and enfeebled articulation of the project vision or a lack of meaningful business impact. One notable example is the infamous Taurus project in the U.K., which failed after spending more than £500 million (Drummond, 1998). The root cause of the failure can be directly attributed to poor executive oversight and ineffective project management techniques that would have otherwise flagged the project well ahead of its projected end date as having a major risk of failure.

This example demonstrates how organizations fail to align their overall strategic goals with the specific objectives of individual projects—clearly a project selection issue that is of relevance to this work. This may also reflect how quickly business strategies evolve in relation to project timelines, especially in mega project settings.

The current literature indicates general agreement among researchers on the differences between leadership and management (e.g., Bennis, 1989; Kotter, 1990; Zaleznik, 1977). There is also an extensive body of literature that has explored this domain previously so exploring these differences has limited utility. It is important to note that, in general, researchers agree that leadership must exert itself most when the business context is vague, dynamic, or challenging. However, there appears to be a lack of citations in the literature indicating how to accomplish this in a project management setting when these same conditions are present.

By definition, project management is also about implementing a change program (Turner et al., 1996; Briner et al., 1996; Cleland, 1999; Turner & Cochrane, 1993) in the form of system changes—as in IT projects—or in building projects, new automotive products, airplanes, or weapons systems. This creates a dilemma for project managers who, when faced with a set of ambiguous circumstances, do not appear to have very many tools at their disposal to address these situations. Again, adding a measurement component may help address what changes should occur and why they are of strategic importance, providing that the strategy itself can be made more measurable.

Another possible leadership issue arises when a corporate culture or a particular internal set of values is incongruous with project success. Again, without measurable project outcomes, it is difficult to both identify and challenge culturally based norms or values that may be barriers to project selection or implementation. This topic is well studied in the abundance of change management literature (Collins & Porras, 1996; Kotter, 1995) and, seemingly, the symptoms and causes of this kind of discord at the corporate level are well understood. More recently, researchers have begun to assess the impact of this topic in a project context (Yukl, 1998), although primarily from a social-psychological perspective.

Since a project manager acts as both a leader and a manager at the same time, and depending on the project and the personalities of the project sponsor(s) and project manager(s) involved, the extent of this overlap (Briner et al., 1996; Cleland, 1999; Morris, 1994) is an important issue within the profession. While important to note, the expectation of this study is that both project managers and sponsors are generally competent in their domains and are able to interact with and understand the issues presented.

As the project management literature shows (i.e., Turner & Muller, 2005) the role that effective leadership plays is widely recognized. For example, Briner et al., (1996) state, "The most significant success factor for project teams is that they have a common and shared idea of what difference they are trying to make as a result of the project" (p. 89). A definition of strategic project outcomes requires exploratory dialogue with project stakeholders, and this requires that organizational leaders have a clear picture of the organization's strategy and link it to these preferred project outcomes. The development of a project's vision is an essential element of the leader's role at the project conceptualization and proposal stage (Christensen & Walker, 2004).

To prevent the loss of a clear project vision, Baccarini (1999) and Davis (1995) offer the Logical Framework Method (LFM) as a method for defining project success. This is important work, as it significantly contributed to the improvement of methods for connecting projects to strategic outcomes. However, the method could be strengthened by linking the LFM to a strategic measurement framework to improve the notion of measurable outcomes. Doing so would enhance the clarity of the team's objectives by implementing its strategy and realizing its projects. It would also enable the method to move from potentially being focused on single projects to multiple projects managed as a portfolio—something it cannot address in its present form. In so doing, organizations could help project teams connect specific project objectives to their current strategic gaps. By linking the outcomes of projects with a measurable vision, organizations could enhance the commitment of the individuals on its project teams to their projects.

Another area of interest is the emerging research on project leadership and the discontinuation or cancellation of a project. For instance, Keil (2000) writes, "Ending runaway projects is one of the toughest executive decisions." He proposes, based on years of accumulated research work on this topic, a four-stage process managers can use to stop the flow of resources to a troubled project and implement an exit strategy. Drummond (1998) provided additional insight into this problem when reporting on the troubled Taurus project in the U.K., and earlier references (Brockner, 1992; Anthes, 1996) note the challenges using their own case studies. All these researchers agree that it is difficult to disengage from a project once it starts and this is problematic. Again, this aspect of the literature would appear to support the need to establish strict methods that do not rely exclusively on executive will or willingness to cancel non-strategic projects; but rather on a systematic way of comparing current projects with proposed projects and determining, at the enterprise level, which ones should be stopped or started on that basis.

What is clear is that better practice implicitly assumes project teams have a clear vision of the project, devolved from a process led by the executive sponsor and/or project leader. This is the process used in project management's traditional triple-constrained model, which focuses on time, budget and quality outcomes and presupposes that all projects that are approved are therefore strategic. What if the projects were not strategic? Or what if the strategy evolves more quickly than the project's timelines? Therefore, it is easy to negate this assumption because organization strategy is so often not expressed in measurable terms. Such strategic ambiguity creates severe leadership challenges and likely makes it impossible for leaders to determine exactly what strategic contribution to expect from any particular project. This creates the decision-making challenge addressed in this book.

FINANCE AND PORTFOLIO THEORY REFERENCES

The origins of PPM lie in the theoretical domain of finance—specifically capital allocation and investment portfolio theory. The basic notion of balancing a portfolio between risk and return is common knowledge and is understood as an overarching objective of sound financial management, both personally and corporately. This was first proposed by Markowitz (1959), and it is a notion for which he was later awarded the Nobel Prize in Economics.

But when we move more deeply into a study of the mechanics of portfolio theory, we find in the associated literature in-depth discussions about how to assess, measure and relate risk and return to assess the true value of a potential activity (Churchman & Ackoff, 1954 for example) before assessing its value. As time goes on, manual calculations are replaced with more substantive mathematical models and model portfolio constructs (Sharpe, 1964; Saaty, Rogers & Pell, 1980) that are often associated with the term "efficient frontier" as the place where return is maximized for any level of acceptable risk. The focus of early writing in project portfolio management often requires a complete economic appraisal of the "fully loaded" costs of a project to compare its anticipated benefits with its costs and risks. In retrospect this seems to be an obvious recommendation, but it was breakthrough thinking at the time. They suggest that companies prioritize those projects that offer the highest likelihood of higher returns, measured capital consumption and a lower probability of risk (something we take as a given in project management methodologies today). In their book *Connecting the Dots* (2003), authors Benko and McFarlan provide a chart that summarizes the comparison between Financial Portfolio Management and Project Portfolio Management as follows:

	Financial Portfolio	Project Portfolio
Assets	Various financial instruments with distinct characteristics.	Various projects with distinct characteristics.
Diversification	Employing multiple financial instruments can reduce risk.	Monitoring project variables—scope, approach, vendors, project managers, etc.—can reduce risk.
Goals	Income and capital gains.	Profitability and growth.
Asset Allocation	Invest according to individual investment goals.	Invest according to overall organizational intentions.
Connections	Correlation.	Interdependency.

Source: Summary Comparison of Portfolio Management Paradigms (Benko & McFarlan, 2003)

While interesting, these discussions are relevant to PPM only to establish that the *a priori* objective of creating and managing a portfolio is always to maximize financial return while minimizing risk. Thus, the optimal portfolio at the efficient frontier is assumed to generate the highest possible return for any given level of risk. In terms of investments, problems can arise because of the inherent risk in the financial instrument itself or as a relationship risk derived from how an instrument or portfolio of instruments relate to each other. Over time, this has led to the basic assumption that risk is minimized through a diversified portfolio. This is known as the assumption of collaborative risk (Maginn & Tuttle, 1990) and it assumes each financial instrument in the portfolio is not inter-dependent and that the choice to include or exclude it can be made without consequences. But this is not true of project settings where interdependencies are a fact.

When we examine the earliest references to project management in this body of literature, the reader should be aware of its specific application in a marketing context (initially in the selection and management of a portfolio of new products or R&D efforts), as in Pessemier & Baker, 1971, where we see references to "program and project decisions" in a research and development context but are still not quite at the stage of complete treatment of all current and proposed projects as a single portfolio. An oft-cited founding reference to PPM is Souder (1973), who, in the article "Utility and Perceived Acceptability of R&D Project Selection Methods," notes that the fundamental issue of project interdependency is distinct from the independent collaboration of financial instruments. Even this title allows the reader to see the early alignment with the financial and mathematical origins of portfolio theory being applied in project management settings. Souder proposes that a more structured model (including mathematical calculus to assess relative risk between projects) would enable corporations to make more informed decisions about which projects to continue and which to stop. Souder followed up in 1975 with a key article in *Management Sciences* titled "Achieving Organizational Consensus with Respect to R&D Project Selection Criteria." In it, he advocates for the use of consistent criteria across both existing and proposed projects, for the purpose of making relative comparisons between them. Thus we begin to see the emergence of PPM in its current incarnation. However, the approach was still considered too complex to be applied by many organizations at the time (Martino, 1995) and so was not often used in practice because of the substantial amount of data and analytical processing required to reach conclusions (Archer & Ghasemzadeh, 1999). It is also clear that the theories on how to combine these disciplines more seamlessly (financial theory, the R&D process and project management) had not yet completely emerged.

As the next two decades unfold, academics continue to be interested in the seductive theoretical simplicity of portfolio theory and it remains a dominant theme in the literature. A development of note is that as desktop computing becomes more readily available, and as the models and approaches become more refined and less complicated to apply, we see more evidence of the in-practice adoption of some of the process recommendations in the literature. Particularly, this can be seen in successful corporate case studies in industries like pharmaceuticals, consumer products and industrial chemicals. Two of many such examples would be Rzasa, Faulkner and Sousa (1990), who explore the application of these techniques in the R&D project selection at Eastman Kodak, and Krumm and Rolle (1992), who looked at the application of decision support and risk analysis at Du Pont. Nearly 20 years after its first emergence in the academic literature, we finally begin to see some practitioner application of the theory in practice.

By 1992, we see continued evidence of the application of capital asset pricing models specifically to project assessment and selection (Khan

& Fiorino, 1992) and interest in how to refine and more accurately forecast and price multi-year investments and returns over time in project settings.

Subsequently, Robert Cooper (a Professor of Marketing at McMaster University) began to evolve the process design combining "stage gates" with the interim assessment of potential risk and return at each stage of the new product development life cycle in order to recommend specific decisions at each stage of a company's project management life cycle (Cooper, 1993). This was considered a practical breakthrough by many practitioners in terms of recommending a sound business process that applies seemingly complex theory in a precise, prescriptive and practical way that organizations could understand and adopt. His full range of work in this area is cited frequently and has been adopted by many corporations around the world. His sound linking of these core disciplines (finance, R&D and project management) has been subsequently reviewed and refined through both practice and additional research on new product R&D and innovation (Khurana & Rosenthal, 1997). Subsequently, Weill and Broadbent (1998) come the closest to any reference found by linking portfolio selections to projects that not only exhibit financial returns but acknowledge the value of strategic returns, such as better information flows, improved business integration, or improvements in quality or customer service. While the article links to the Balanced Scorecard concept, it does not directly articulate the link between project scoring criteria and the emergent strategy measures. However, there is some early evidence emerging of a desire by practitioners to now establish clear links between project portfolio selection practices and business strategy intentions, if not the measures themselves (Benko & McFarlan, 2003; Artto, 2001).

One thing remains common to all of these previous citations: they are primarily based on making more profit by maximizing return while minimizing risk within a private sector context. Less has been written about how to apply this theory where this critical assumption is not present, as is the case in the not-for-profit and public sectors.

BSC AND PERFORMANCE MANAGEMENT REFERENCES

An extensive amount of work has appeared in the literature related to the application of the Balanced Scorecard in corporate settings. Much of this literature stems from the original work of Drs. Kaplan & Norton (1993; 1996; 1998; 2004), who defined a multi-dimensional framework that translates an organization's strategy into specific, measurable objectives around four specific dimensions (financial, customer, internal/operational and innovation & learning). The measures associated with each objective provide a "dashboard" or "scorecard" of the organization's progress towards its objectives over time. The concepts are simple in theory, and their value in practice is clear, so it is surprising how many organizations fail to implement this strategy properly.

Subsequent authors have built on this original work (for instance, the "Success Dimensions" framework by Shenhar & Dvir, 1996 and the Dynamic Multi-dimensional Performance (DMP) framework by Maltz, Shenhar & Reilly, 2003). The authors in most cases simply extend the methodology to specific settings (i.e., addressing the more rapid rate of business change in technology firms, as Shenhar & Dvir suggest) or provide suggested enhancements to the original methodology to repair perceived gaps (such as the emphasis, in the DMP framework, on other "soft" factors and its incorporation of environmental variables of performance).

Beyond the original articles and books defining the BSC methodology itself, other researchers have explored the issue of balanced performance measurement. A review of 51 empirical studies of entrepreneurial firms published between 1987 and 1993 reveals that most firms use only financial measures to gage their success (Murphy, Trailer & Hill, 1996) and, not surprisingly, the most common performance measures used also related to efficiency, revenue growth and profit. However, the use of any single dimension (i.e., finance) as a surrogate for overall organizational performance can produce a false result. Chakravarthy (1986) used the firms noted by Peters & Waterman (1982) in their book *In Search of Excellence* as "excellent" and used classic financial measures (ROE, ROC, ROS) to attempt to correlate the performance of these firms with their financial results. He concluded that these measures were too narrow an interpretation of performance, were incapable of distinguishing future differences in performance among the firms and only reported on historical performance. This conclusion directly supports the notion of balanced performance management as the superior method for measuring strategic accomplishment, as the BSC purports to do.

While some limitations of the BSC have been identified (for example, Atkinson, Waterhouse & Wells, 1997; Smith, 1998), the critiques are relatively minor when compared to the number of organizations successfully using the BSC and these limitations do not impair its overall intended purpose: clarifying firm strategy.

Therefore, it is important to demonstrate that others, both prior to and after Kaplan and Norton, have done notable work in this same genre and domain (Eccles, 1991; Sveiby, 1997; Nealy, 2002). In some instances, they have proposed modifications or variations to the existing notions of balanced performance management frameworks or provided additional examples of their application in specific settings. However, generally all experts in the field support the notion that if measurement in a corporate setting is to be effective it must be multi-dimensional and represent both the tangible and intangible components of organizational strategy (Nealy, 2002). It is not important for a corporate performance system to be perfect, only to be present, in my view.

At the outset, the primary focus of the balanced scorecard was its application in the private sector; in fact, with an emphasis on large, F-1000 U.S. companies initially. Over time, however, the methodology has established itself as having equal relevance in the public and not-for-profit sectors (albeit with slight modifications to the application of measures in the financial domain because of the absence of a profit motive) with a more recent book by Kaplan and Norton (2004) citing several case studies from these two sectors. Equally, there is an emerging body of work by those who are attempting to specialize in the application of the BSC in government (Whittaker, 2002). The fact that other researchers feel the BSC is both applicable to and useful for the public sector practitioner is an important endorsement.

The recommended modifications by these authors to the original Kaplan and Norton methods seem generally useful and improve the usability of the BSC methodology in the public and not-for-profit sectors. As a result, some practitioners, myself included, have incorporated these adjustments into practice. Again, of ultimate importance is the clear supporting conclusion that the public sector, like the private sector, needs to align its measurement with strategy. The benefits of applying this approach at the organization level have been clearly substantiated and should be noted as best practice by leaders in this sector.

While proven at the enterprise level of both the private and public sectors, less has been written about connecting the balanced scorecard to project management methodologies. Recently, the beginnings of interest in this topic (Stewart, 2001; Stewart & Mohamed, 2001) have emerged. And some researchers have attempted to apply it to tangential areas such as IT service level management or service level agreements, including for specific project demands (Van Grembergen et al., 2003). This would seem to support the value of the basic methodology as a strategic measurement tool that can be applied in new ways. However, the overall lack of citations may be due to the rather recent nature of the BSC itself and the natural inclination of researchers to go from the macro to the micro level of any new concept over time. It is interesting that even more recent citations around PPM (for instance, Artto et al., 2001; Morris & Jamieson, 2004) do not directly touch on the BSC and only obliquely refer to the issue of strategic measurement. Perhaps as more studies begin to further explore the mechanics of applying the BSC in new contexts, their findings will spur other researchers to focus on this area.

Of some note is that these studies primarily apply to IT and managing individual IT-related projects and do not apply to the general domain of enterprise project management or portfolio project management. Results from one earlier study (Hersey et al., 1996) suggest that a project level BSC can become a tool that provides an indirect form of influence on daily decision-making within a project team. And Norrie and Walker (2004) establish this tool as having a powerful influence on project outcomes—perhaps more so than other methods of influence in terms of accelerating project outcomes. But generally, the limited nature of literature in this area supports the importance of this work as it attempts to link the use of balanced performance management techniques like the BSC with more advanced project management methodologies and, particularly, with project portfolio selection.

BIBLIOGRAPHY

Amit, R. and P.J.H. Schoemaker (1993). "Strategic assets and organizational rent." *Strategic Management Journal* 14(1): 33–46.

Andriessen, D. and R. Tissen (2000). *Weightless wealth*. Harlow, Great Britain, Pearson Education Limited.

Anthes, G. (1996). "Users fall short on 'net security planning.'" *Computerword* 30(29): 8.

Archer, N.P. and F. Ghasemzadeh (1998). "A decision support system for project portfolio selection." *International Journal of Technology Management* 16(1/2/3): 105–114.

Archer, N.P. and F. Ghasemzadeh (1999). "An integrated framework for project portfolio selection." *International Journal of Project Management* 17(4): 207–216.

Archer, N.P. and F. Ghasemzadeh (2004). *Project portfolio selection and management. The Wiley Guide to Managing Projects*. P. W. G. Morris and J. K. Pinto. New York, NY, John Wiley and Sons, Inc.: 237–255.

Archibald, R. (2003). *Managing high technology programs and projects*. New York, John Wiley & Sons.

Archibald, R.D. (2003). "Part 1: Project management within organizations." *State of the Art of Project Management*.

Archibald, R.D. (2003). "Part 4: Project management in the next five years." *State of the Art of Project Management*.

Argyris, C. (1977). "Double loop learning in organizations." *Harvard Business Review* 55(5): 115–125.

Artto, K.A., M. Martinsuo, et al. (2001). *Project portfolio management: strategic management through projects*. Helsinki, Finland, Project Management Association.

Ashurst, C. and N.F. Doherty (2003). "Towards the formulation of a 'best practice' framework for benefits realisation in IT projects." *Electronic Journal of Information Systems Evaluation* 6(2).

Atkinson, A. and J.Q. McCrindell (1994). "A new perspective on control in government (part four)." *CMA—Management Accounting Manager* 68(6): 28.

Atkinson, A.A., J.H. Waterhouse, et al. (1997). "A stakeholder approach to strategic performance measurement." *Sloan Management Review* (Spring): 25–37.

Avison, D. and Francis Lau et al. (1999). "Action Research." *Communications of the ACM:* 42 (1): 94–98

Avison, D.E., R. Baskerville, et al. (2001). "Controlling action research projects." *Information, Technology and People* 14(1): 28–45.

Ayres, R. and D. Russell (2001). "The Commonwealth's outcomes and outputs framework: implications for management accountability." *Canberra Bulletin of Public Administration*(99): 33–39.

Baburoglu, O.N. and I. Ravn (1992). "Normative action research." *Organization Studies* 13(1): 19–34.

Baccarani, D. (1999). "The logical framework method for defining project success." *Project Management Journal* 30(4): 25–32.

Bakan, J. (2004). *The corporation: the pathological pursuit of profit and power.* Toronto, ON, Viking.

Ballow, J.J., R. Burgman, et al. (2004). "Managing for shareholder value: intangibles, future value and investment decisions." *The Journal of Business Strategy* 25(3): 26–34.

Barber, M. (2002). *Delivering on the promises.* Going Public, Victorian Division of the Institute of Public Administration Australia: 12–17.

Baron, D.P. (1995). "The nonmarket strategy system." *Sloan Management Review* (Fall 1995): 73–85.

Bartlett, C.A. and S. Ghoshal (1995). "Changing the role of top management: beyond structure to process." *Harvard Business Review* (May-June): 132–142.

Baskerville, R. (1999). "Investigating information systems with action research." *Communications of the AIS* 2(19): 1–32.

Baskerville, R. and J. Pries-Heje (1999). "Grounded action research: a method for understanding IT in practice." *Accounting Management and Information Technology* 9(1): 1–23.

Baskerville, R. and A.T. Wood-Harper (1996). "A critical perspective on action research as a method for information systems research." *Journal of Informational Technology* 11(3): 235–246.

Baskerville, R. and A.T. Wood-Harper (1998). "Diversity in information systems action research methods." *European Journal of Information Systems* 7(2): 90–107.

Belassi, W. and O.I. Turkel (1996). "A new framework for determining critical success/failure factors in projects." *International Journal of Project Management* 14(3): 141–151.

Belout, A. (1998). "Effects of human resource management on project effectiveness and success: toward a new conceptual framework." *International Journal of Project Management* 16(1): 21–26.

Benbasat, I. and R.W. Zmud (1999). "Empirical research in information systems: the practice of relevance." *MIS Quarterly* 23(1): 3–16.

Benko, C. and F.W. McFarlan (2003). *Connecting the dots: aligning projects with objectives in unpredictable times*. Boston, MA, Harvard Business School Press.

Bennington, P. and D. Baccarini (2004). "Project benefits management in IT projects: an Australian perspective." *Project Management Journal* 35(2): 20–30.

Bennis, W. (1989). *On becoming a leader*. San Franciso, CA, Perseus Publishing.

Bennis, W. and B. Nanus (1997). *Leaders: strategies for taking charge*. New York, NY, Harper Business.

Bennis, W., G. Spevietzer, et al. (2001). *The future of leadership*. San Francisco, CA, Jossey-Bass.

Blair, M.M. and S.M.H. Wallman (2001). *Unseen wealth: report of the Brookings Task Force on intangibles*. Washington DC, Brookings Institution Press.

Block, T.R. (1999). "The seven secrets of a successful project office." PM Network.

Bourne, L. and D.H.T. Walker (2005). "The paradox of control." *Team Performance Management* 11(5/6): 157–178.

Boynton, A.C., G.C. Jacobs, et al. (1995). "Whose responsibility is IT management?" *Sloan Management Review* 33(4): 32–38.

Bredillet, C.N. (2002). "Proposition of a systemic and dynamic model to design lifelong learning structure: the quest of the missing link between men, team, and organizational learning." *The Frontiers of Project Management Research*. D.P. Slevin, D.I. Cleland and J.K. Pinto. Newtown Square, Project Management Institute: 73–95.

Brigman, L. (2004). "Evolving strategy execution at Revcor." *Strategic HR Review* 3(6): 32–35.

Briner, W., C. Hastings, et al. (1996). *Project leadership*. Aldershot, U.K., Gower.

Brockner, J. (1992). "The escalation of commitment to a failing course of action: toward theoretical progress." *The Academy of Management Review* 17(1): 39–61.

Brown, K. and J. Waterhouse (2003). "Change management practices: is a hybrid model a better alternative for public sector agencies?" *The International Journal of Public Sector Management* 16(3): 230–241.

Bryde, D.J. (2003). "Modelling project management performance." *International Journal of Quality and Reliability Management* 20(2/3): 229–254.

Buckmaster, N. (1999). "Associations between outcome measurement, accountability and learning for non-profit organisations." *The International Journal of Public Sector Management* 12(2): 186–197.

Busby, J.S. (1999). "An assessment of post-project reviews." *Project Management Journal* 30(3): 23–29.

Byers, C.R. and D. Blume (1994). "Tying critical success factors to systems development." *Information and Management* 26(1): 51–61.

Campbell, A. and M. Alexander (1997). "What's wrong with strategy?" *Harvard Business Review* 75(6): 42–51.

Canada, J.R. and J. White (1980). *Capital investment analysis for management and engineering.* Englewood Cliffs, NJ, Prentice-Hall, Inc.

Carr, N.G. (2003). "IT doesn't matter." *Harvard Business Review* (5): 5–12.

Carroll, T. (2003). "Delivering business benefits from projects: dovetailing business and IT: a case study from Standard Chartered Bank," Standard Chartered Bank.

Caupin, G., H. Knopfel, et al., Eds. (1999). *IPMA Competence Baseline (ICB) Version 2.0.* Bremen, Germany, International Project Management Association.

Chakravarthy, B.S. (1986). "Measuring strategic performance." *Strategic Management Journal* 7(5): 437–458.

Chan, Y.E. and S. Huff (1993). "Strategic information systems alignment." *Business Quarterly* 58(1): 51–54.

Checkland, P. (1981). *Systems thinking, systems practice.* Chichester, U.K., Wiley and Sons, Inc.

Checkland, P. (1991). "From framework through experience to learning: the essential nature of action research." *Information Systems Research: Contemporary Approaches and Emergent Traditions.* H. E. Nissen, H. K. Klein and R. Hirschheim. Amsterdam, The Netherlands, Elsevier North-Holland, Inc.: 397–403.

Checkland, P. and S. Holwell (1998). "Action research: its nature and validity." *Systemic Practice and Action Research* 11(1): 9–21.

Chisholm, R.F. and M. Elden (1993). "Features of emerging action research." *Human Relations* 46(2): 275–298.

Choo, C.W. (1997). *The knowing organization: how organizations use information to construct meaning, create knowledge and make decisions.* New York, NY, Oxford University Press.

Christensen, D. and D.H.T. Walker (2004). "Understanding the role of vision in project success." *Project Management Journal* 35(3): 39–52.

Churchman, W.C. and R.L. Ackoff (1954). "An approximate measure of value." *Journal of Operations Research Society of America* 2(2): 172–187.

Ciborra, Claudio U. et al. (2000). *From Control to Drift: The Dynamics of Corporate Information Infrastructure.* London, U.K., Oxford University Press.

Cicmil, S. (1997). "Critical factors in effective project management." *The TQM Magazine* 9(6): 390–396.

Clarke, A. (1999). "A practical use of key success factors to improve the effectiveness of project management." *International Journal of Project Management* 17(3): 139–145.

Cleland, D.I. (1999). *Project management: strategic design and implementation.* Singapore, Asia, McGraw-Hill Education.

Cleland, D.I. (2004). "Strategic management: the project linkages." *The Wiley guide to managing projects.* P.W.G. Morris and J.K. Pinto. Hoboken, New Jersey, John Wiley & Sons, Inc.: 206–222.

Collins, D. (1990). "Management by subjective." *Executive Development* 3(3): 14–15.

Collins, J. and J.I. Porras (1996). "Building your company's vision." *Harvard Business Review* 74(5): 65–78.

Cooke-Davies, T. (2001). Managing benefits: the key to project success. *Project Manager Today*: 1–3.

Cooke-Davies, T. (2004). "Project success." *The Wiley guide to managing projects.* P.W.G. Morris and J.K. Pinto. Hoboken, NJ, John Wiley & Sons Inc: 99–122.

Cooke-Davies, T. J. (2000). *Towards improved project management practice: uncovering the evidence for effective practices through empirical research.* Leeds, U.K., Leeds Metropolitan University.

Cooke-Davies, T. (2002). "The "real" success factors on projects." *International Journal of Project Management* 20: 185–190.

Cooper, R. (1993). *Winning at new products.* Reading, MA, Addison-Wesley.

Cooper, R. (1997). "Portfolio management in new product development: lessons from leaders II." *Research Technology Management* 40(6): 43–52.

Cooper, R., S. Edgett, et al. (1997). "Portfolio management in new product development: lessons from leaders I." *Research Technology Management* 40(5): 16–28.

Cooper, R., S. Edgett, et al. (1998). *Portfolio management for new products.* Reading, PA, Perseus Books.

Crawford, J.K. (2001). Portfolio management: overview and best practices. *Project management for business professionals: a comprehensive guide.* J. Knutson. New York, NY, John Wiley & Sons, Inc.: 33–48.

Crawford, J.K. (2002). *The strategic project office.* New York, NY, Marcel Dekker AG.

Crawford, L. (2005). "Senior management perceptions of project management competence." *International Journal of Project Management* 23(1):7–16.

Cusick, K. (2001). *How to use capability maturity models to help manage projects effectively.* Project Management Institute Professional Seminar, Annual Seminars and Symposia, Philadelphia, PA.

Dai, C.X. and William Wells (2004). "An exploration of project management office features and their relationship to project performance." *International Journal of Project Management* 22(4): 523–532.

Darke, P., G.G. Shanks, et al. (1998). "Successfully completing case study research: combining rigour, relevance and pragmatism." *Information Systems Journal* 8(4): 273–290.

Davis, K.H. (1995). Logical framework analysis: a methodology to turn vision into reality. AIPN National Conference, Adelaide, Australia.

Davison, R.M., M.G. Martinsons, et al. (2004). "Principals of canonical action research." *Information Systems Journal* 14(1): 65–86.

De Maio, A., R. Verganti, et al. (1994). "A multi-project management framework for new product development." *European Journal of Operational Research* 78(2): 178–191.

de Wit, A. (1988). "Measurement of project success." *International Journal of Project Management* 6(3): 164–170.

Denzin, N.K., S. Yvonna, et al., Eds. (2000). *The handbook of qualitative research.* London, UK, Sage Publications.

Dickens, L. and K. Watkins (1999). "Action research: rethinking Lewin." *Management Learning* 30(2): 127–140.

Dinsmore, P.C. (1998). "How grown-up is your organization?" *PM Network* 12 (6): 24–26.

Dinsmore, P.C. (1999). *Winning in business with enterprise project management.* New York, NY, AMACOM.

Dixon, M., Ed. (2000). *Project management body of knowledge.* High Wycombe, UK, Association for Project Management.

Donnelly, M. (1999). "Making the difference: quality strategy in the public sector." *Managing Service Quality* 9(1): 47–52.

Drummond, H. (1998). "Riding a tiger: some lessons of Taurus." *Management Decision* 36(3): 141–146.

Dvir, D., S. Lipovetsky, et al. (1998). "In search of project classification: a non-universal approach to project success factors." *Research Policy* 27(9): 915–935.

Dvir, D., E. Segev, et al. (1993). "Technology's varying impact on the strategic success of business units within the Miles and Snow typology." *Strategic Management Journal* 14(2): 155–162.

Dye, L.D. and J.S. Pennypacker, Eds. (1999). *Project portfolio management: selecting and prioritizing projects for competitive advantage.* West Chester, PA., Centre for Business Practices.

Eagle, K. (2004). "Translating strategy into results: the origins and evolution of Charlotte's Corporate Scorecard." *Government Finance Review* 20(5): 16–27.

Earl, M.J. (1993). "Experiences in strategic information systems planning." *MIS Quarterly* 17(1): 1–24.

Eccles, R.G. (1991). "The performance measurement manifesto." *Harvard Business Review* 69(1): 131–137.

Eden, C. and C. Huxham (1996). "Action research for management research." *British Journal of Management*, 7(1): 75–86.

Ejigiri, D.D. (1994). "A generic framework for programme management: the cases of Robert Moses and Miles Mahoney in the U.S." *International Journal of Public Sector Management* 7(1): 53–66.

Elden, M. and R. F. Chisholm (1993). "Emerging varieties of action research: introduction to the special issue." *Human Relations* 46(2): 121–142.

Elkington, J. (1997). *Cannibals with forks.* London, U.K., Capstone Publishing.

Elton, J. and J. Roe (1998). "Bringing discipline to project management." *Harvard Business Review* (March–April).

Feldman, J. (2003). "Lessons from the field: beyond ROI." *Network Computing* 14(4): 34–41.

Ferlie, E. and P. Steane (2002). "Changing developments in NPM." *International Journal of Public Administration* 25(12): 1459–1460.

Fitzgerald, B. and D. Howcroft (1998). "Towards dissolution of the IS research debate: from polarization to polarity." *Journal of Informational Technology* 13(4): 313–326.

Forsberg, K., H. Mooz, et al. (1996). *Visualizing project management*. New York, NY, Wiley.

Foti, R. (2003). "Make your case." *PM Network* 17: 36–43.

Frame, J.D. (1994). *The new project management*. San Francisco, CA, Jossey-Bass.

Frame, J.D. (2003). *Managing projects in organizations: how to make the best use of time, techniques and people*. San Francisco, Jossey-Bass.

Future and Innovation Unit (2001). *Creating value from your intangible assets*. London, U.K., Department of Trade and Industry (U.K.): 38.

Gadiesh, O. and J.L. Gilbert (2001). "Transforming corner-office strategy into frontline action." *Harvard Business Review* 76(5): 72–79.

Garcia-Ayuso, M. (2003). "Intangibles: lessons from the past and a look into the future." *Journal of Intellectual Capital* 4(4): 597–604.

Gedansky, L.M. (2002). "Inspiring the direction of the profession." *Project Management Journal* 33(1): 4.

Germonprez, M. and L. Mathiassen (2004). *The role of conventional research methods in information systems research*. IFIP 8.2, Manchester, U.K.

Gioia, J. (1996). "Twelve reasons why programs fail." *PM Network* 10(11): 16–20.

Greiner, L.E. and V.E. Schein (1988). *Power and organization development: mobilizing power to implement change*. Reading, MA, Addison-Wesley.

Grundy, T. (1998). "Strategy implementation and project management." *International Journal of Project Management* 16(1): 43–50.

Hall, M.-J. (2002). "Aligning the organisation to increase performance results." *The Public Manager* (Summer 2002): 7–10.

Hall, M. and R. Holt (2002). "U.K. public sector project management: a cultural perspective." *Public Performance and Management Review* 25(3): 298–312.

Hamel, G. (1995). "Reinventing the company." *Executive Excellence* 12(10): 9–13.

Hamel, G. (1996). "Strategy as revolution." *Harvard Business Review* 74(4): 69–82.

Hamel, G. and C. K. Prahalad (1994). *Competing for the future*. Boston, MA, Harvard Business School Press.

Hansen, M.T., Nohria, N., and T. Tierney (1999). "What's your strategy for managing knowledge?" *Harvard Business Review* (March-April): 106–116

Hax, A. and V. Majluf (1996). *The strategic concept and process: a pragmatic approach*. Upper Saddle River, NJ, Prentice Hall.

Hendriks, M., B. Voeten, et al. (1999). "Human resource allocation in a multi-project environment." *International Journal of Project Management* 17(3): 181–188.

Henderson, J.C. and N. Venkatraman (1993). "Strategic alignment: Leveraging information technology for transforming organizations." *IBM Systems Journal* 32(1): 4–16

Hersey, P., K. Blanchard, et al. (1996). *Management of organizational behavior: utilizing human resources*. London, U.K., Prentice Hall International.

Hirschheim, R. and H.K. Klein (1994). "Realizing emancipatory principles in information systems development: the case for ETHICS." *MIS Quarterly* 18(1): 83–102.

HM Treasury, U. (2003). "The Green Book: appraisal and evaluation in central government." Stationery Office Books, 2003.

Hobbs, B. and M. Aubry (2006). "Identifying the Structure that Underlies the Extreme Variety Found Among PMO's." PMI Research Conference Paper, Montreal, PQ: 2006.

Hoenig, C. (2003). "Hidden assets: strategies for managing your intangible leadership capital." *CIO Magazine* (May 2003): 36–38.

Hodgson, D. (2002). "Disciplining the Professional: The Case of Project Management." *Journal of Management Studies* 39(6): 803–821.

Hussi, T. and G. Ahonen (2002). "Managing intangible assets: a question of integration and balance." *Journal of Intellectual Capital* 3(3): 277–286.

Ibbs, C.W. and Y.H. Kwak (1997). *Benefits of project management: financial and organisational rewards to corporations.* Upper Darby, PA, Project Management Institute.

Jamieson, A. and P.W.G. Morris (2004). "Moving from corporate strategy to project strategy. The Wiley guide to managing projects." P. W. G. Morris and J. K. Pinto. Hoboken, NJ, John Wiley & Sons, Inc.: 177–205.

Jin, K.G. (2000). "Power-based arbitrary decisional actions in the resolution of MIS project issues: a project manager's action research perspective." *Systemic Practice and Action Research* 13(3): 345–390.

Jugdev, K. and J. Thomas (2002a). Blueprint for value creation: developing and sustaining a project management competitive advantage through the resource based view. Project Management Institute Research Conference 2002: Frontiers of Project Management Research and Application. Seattle, WA, PMI.

Jugdev, K. and J. Thomas (2002b). "From operational process to strategic asset: the evolution of project management's value in organizations." Project Management Institute 33rd Annual Symposium and Conference. San Antonio, TX.

Kaplan, R.S. and N. Klein (1995). *Harvard Business School Case 9-195-210: Chemical Bank: implementing the balanced scorecard.* Boston, MA, Harvard Business School.

Kaplan, R.S. and J.A. Maxwell (1994). "Qualitative research methods for evaluating computer information systems." *Evaluating health care information systems: approaches and applications.* J. G. Anderson, C. E. Aydin and S. J. Jay. Thousand Oaks, CA, Sage.

Kaplan, R.S. and D.P. Norton (1992). "The balanced scorecard: measures that drive performance." *Harvard Business Review* 70(1): 71–79.

Kaplan, R.S. and D.P. Norton (1993). "Putting the balanced scorecard to work." *Harvard Business Review* 71(5): 134–142.

Kaplan, R.S. and D.P. Norton (1996). "Using the balanced scorecard as a strategic management system." *Harvard Business Review* 74(1): 75–85.

Kaplan, R.S. and D.P. Norton (1998). "Using the balanced scorecard as a strategic management system." *Harvard Business Review on Measuring Corporate Performance*: 183–211.

Kaplan, R.S. and D.P. Norton (2004). "Measuring the strategic readiness of intangible assets." *Harvard Business Review*(2): 52–63.

Kaplan, R.S. and D.P. Norton (2004). *Strategy maps: converting intangible assets into tangible outcomes*. Boston, MA, Harvard Business School Publishing Corporation.

Katzenback, J.R. and D.K. Smith (1993). "The discipline of teams." *Harvard Business Review* 71(2): 111–120.

Kearns, K.P. (2000). *Private sector strategies for social sector success: the guide to strategy and planning for public and non-profit organizations*. San Francisco, Jossey-Bass.

Keen, J. (2003). "Don't ignore the intangibles." *CIO Magazine* (October 2003): 20–22.

Keen, J. (2003). "Plugging leaky business cases." *CIO Magazine* (May 2003): 25–26.

Keen, J. and B. Digrius (2003). "The emotional enigma of intangibles." *CIO Magazine* (April): 104–107.

Keil, M. (2000). "An investigation of risk perception and risk propensity on the decision to continue software development project." *Journal of Systems and Software* 53(2): 145–157.

Kemmis, S. and R. McTaggart (1988). *The action research planner*. Victoria, Australia, Deakin University Press.

Kerzner, H. (2000). *Applied project management: best practices on implementation*. New York, NY, Wiley & Sons, Inc.

Kerzner, H. (2006). *Project Management: A Systems Approach to Planning, Scheduling, and Controlling* (9th Edition). New York, NY, Wiley & Sons, Inc.

Khan, A.M. and D.P. Fiorino (1992). "The Capital Asset Pricing Model in project selection: a case study." *Engineering Economist* 37(2): 145–159.

Khurana, A. and S.R. Rosenthal (1997). "Integrating the fuzzy front end of new product development." *Sloan Management Review* 38(2): 103–120.

Kiernan, M.J. (1995). *Get innovative or get dead*. Toronto, ON, Douglas and McIntyre.

King, M. and L. McAulay (1997). "Information technology investment evaluation: evidence and interpretations." *Journal of Information Technology* 12: 131–143.

Kippenberger, T. (2000). "Management's role in project failure." *The Antidote* (27).

Kippenberger, T. (2000). "Managing the business benefits." *The Antidote* (27): 28–29.

Kira, D.S., M.I. Kusy, et al. (1990). "A specific decision support system (SDSS) to develop an optimal project portfolio mix under uncertainty." *IEEE Transactions on Engineering Management* 37(3): 213–221.

Klein, H.K. and M.D. Myers (1999). "A set of principles for conducting and evaluating interpretive field studies in information systems." *Management Information Systems Quarterly* 23(1): 67–94.

Kloppenborg, T.J. and W.A. Opfer (2002). "The current state of project management research: trends, interpretations, and predictions." *Project Management Journal* 33(2): 5–18.

Kock, N., D. Avison, et al. (1999). IS action research: can we serve two masters? 20th International Conference on Information Systems., New York, NY, The Association for Computing Machinery.

Kock, N.F., R.J. McQueen, et al. (1997). "Can action research be made more rigorous in a positivistic sense? the contribution of an interpretive approach." *Journal of Systems and Information Technology* 1(1): 1–24.

Kotter, J.P. (1990). *A force for change: how leadership differs from management.* New York, NY, Free Press.

Kotter, J.P. (1995). "Leading change: why transformation efforts fail." *Harvard Business Review* 73(2): 59–67.

Kotter, J.P. (1999). *John P. Kotter on what leaders really do.* Boston, MA, Harvard Business School Press.

KPMG (2004). "KPMG's International 2002-2003 Programme Management Survey: why keep punishing your bottom line?" Sydney, Australia, *KPMG Information Risk Management*: 19.

Krumm, F. and C. Rolle (1992). "Management and application of decision and risk analysis in DuPont." *Interfaces* 22(6): 84–93.

Laszlo, G.P. (1999). "Project management: a quality management approach." *The TQM Magazine* 11(3): 157–160.

Leatherman, S., D. Berwick, et al. (2003). "The business case for quality: case studies and an analysis." *Health Affairs* 22(2): 17.

Lee-Kelley, L. (2002). "Situational leadership: managing the virtual project team." *Journal of Management Development* 21(6): 461–476.

Lee, A.S. (1999). "Rigor and relevance in MIS research: beyond the approach of positivism alone." *MIS Quarterly* 23(1): 29–22.

Lev, B. (2001). *Intangibles: management, measurement and reporting.* Washington DC, The Brookings Institution.

Levene, R.J. and A. Braganza (1996). "Controlling the work scope in organisational transformation: a programme management approach." *International Journal of Project Management* 14(6): 331–340.

Levin, M. (1994). "Action research and critical systems thinking: two icons carved out of the same log?" *Systems Practice* 7(1): 25–41.

Lewicki, R.J., D.J. McAllister, et al. (1998). "Trust and distrust: relationships and realities." *The Academy of Management Review* 23(3): 438–458.

Liebs, S. (1992). "We're all in this together." *Information Week.*

Lin, C.Y. (2002). "An investigation of the process of IS/IT investment evaluation and benefits realisation in large Australian organisations." School of Information Systems. Perth, Australia, Curtin University of Technology.

Llewellyn, S. and E. Tappin (2003). "Strategy in the public sector: management in the wilderness." *Journal of Management Studies* 40(4): 955–982.

Longman, A. and J. Mullins (2004). "Project management: key tool for implementing strategy." *Journal of Business Strategy* 25(5): 54–60.

Loosemore, M. (1999). "Responsibility, power and construction conflict." *Construction Management and Economics* 17(6): 699–709.

Lovell, C.A.K. (1993). "Production frontiers and productive efficiency." *The Measurement of Productive Efficiency: Techniques and Applications.* H.O. Fried, C.A.K. Lovell and S.S. Schmidt. New York, NY, Oxford University Press.

Low, J. and P.C. Kalafut (2002). *Invisible advantage: how intangibles are driving business performance.* Cambridge, MA, Perseus Publishing.

Luftman, J. and T. Briner (1999). "Achieving and sustaining business–IT alignment." *California Management Review* 42(1): 109–122.

Luftman, J.N., Ed. (1996). *Aligning Business and IT strategies.* New York, NY, Oxford University Press.

Lyytinen, K. (1999). "Empirical research in information systems: on the relevance of practice in thinking of IS research." *MIS Quarterly* 23(1): 25–27.

Maginn, J.L. and D.L. Tuttle, Eds. (1990). *Managing investment portfolios: a dynamic process.* Boston, MA, Warren Gorham & Lamont.

Maltz, A.C., A.J. Shenhar, et al. (2003). "Beyond the balanced scorecard: refining the search for organizational success measures." *Long Range Planning* 36(2): 187–204.

Markowitz, H. (1959). *Portfolio selection: efficient diversification of investments.* New Haven, CT, Yale University Press.

Marsh, I. (1999). "Program strategy and coalition building as facets of new public management." *Australian Journal of Public Administration* 58(4): 54–67.

Martin, R.A. (1993). "Changing the mind of the corporation." *Harvard Business Review* 71(6): 81–94.

Martino, J.P. (1995). *Research and development in project selection.* New York, NY, Wiley.

Martinsuo, M. and P. Dietrich (2002). "Public sector requirements towards project portfolio management." *Frontiers of Project Management Research and Application—Proceedings of PMI Research Conference 2002*, Seattle, WA, Project Management Institute.

Mathiassen, L. (1998). "Reflective systems development." *Scandinavian Journal of Information Systems* 10(1&2): 67–118.

Mathiassen, L. (2002). "Collaborative practice research." *Information, Technology & People* 15(4): 321–345.

McKay, J. and P. Marshall (2001). "The dual imperatives of action research." *Information Technology & People* 14(1): 46–59.

McLaughlin, J. (2004). "Winning project approval: writing a convincing business case for project funding." *Journal of Facilities Management* 2(4): 330–337.

McLean, E.R. and J.V. Soden (1977). *Strategic planning in MIS.* New York, NY, Wiley.

McNally, S.J. (2000). "The taste of victory: thoughts on successful project management." *Pennsylvania CPA Journal* 71(3): 14–15.

McNiff, J. and J. Whitehead (2000). *Action research in organizations.* London, U.K., Routledge.

McTaggart, R. (1991). "Principles for participatory action research." *Adult Education Quarterly* 41(3): 168–187.

Merriam-Webster. (2005). "Merriam-Webster Online." Retrieved 14 January, 2005, from www.m-w.com.

Meyer, J.P. and N.J. Allen (1991). "A three component conceptualization of organizational commitment." *Human Resource Management Review* 1: 61–89.

Miles, M.B. and A.M. Huberman (1994). *Qualitative data analysis: an expanded sourcebook.* Thousand Oaks, CA, Sage.

Miller, G.J. (1989). "Unique public sector strategies." *Public Productivity and Management Review* 13(2): 133–144.

Mingers, J. (2001). "Combing IS research methods: towards a pluralist methodology." *Information Systems Research* 12: 240–249.

Mintzberg, H. (1994). *The Rise and Fall of Strategic Planning.* New York, NY, Prentice-Hall.

Mintzberg, H., B. Ahlstrand, et al. (1998). *Strategic safari: a guided tour through the wilds of strategic management.* London, U.K., Pearson Education Press.

Mintzberg, H. and J.A. Waters (1985). "Of strategies, deliberate and emergent." *Strategic Management Journal* 6(3): 257–272.

Montealegre, R. and M. Keil (2000). "De-escalating information technology projects: lessons from the Denver International Airport." *MIS Quarterly* 23(3): 417–447.

Morris, P.W.G. (1988). *The anatomy of major projects.* New York, NY, Wiley & Sons.

Morris, P.W.G. (1994). *The management of projects.* London, U.K., Thomas Telford Services Ltd.

Morris, P.W.G. (1997). *The management of projects.* 2nd Edition. London, U.K., Thomas Telford.

Morris, P.W.G. (2000). "Research into revising the APM project management body of knowledge." *International Journal of Project Management* 18(1): 155–164

Morris, P.W.G. (2002). "Research trends in the 1990s: the need now to focus on the business benefit of project management." *The Frontiers of Project Management.* D.P. Slevin, D.I. Cleland and J.K. Pinto. Newtown Square, Project Management Institute: 31–56.

Morris, P.W.G. (2004). "Moving from corporate strategy to project strategy: leadership in project management." *PMI Research Conference 2004*, London, U.K., Project Management Institute.

Morris, P.W.G. and A. Jamieson (2004). *Translating corporate strategy into project strategy.* Atlanta, GA, Project Mangement Institute.

Morris, P.W.G. and A. Jamieson (2005). "Moving from corporate strategy to project strategy." *Project Management Journal* 36(4): 5–18.

Morris, P.W.G. and J.K. Pinto, Eds. (2004). *The Wiley guide to managing projects.* New York, NY, John Wiley & Sons.

Moss-Kanter, R. (1990). *When giants learn to dance*. New York, NY, The Free Press.

Moss-Kanter, R. (1992). *The challenge of organizational change*. New York, NY, The Free Press.

Mumford, E. (2001). "Advice for an action researcher." *Information Technology & People* 14(1): 12–27.

Munns, A. K. and B. F. Bjeirmi (1996). "The role of project management in achieving project success." *International Journal of Project Management* 14(2): 81–88.

Murphy, G. B., J. W. Trailer, et al. (1996). "Measuring performance in entrepreneurship research." *Journal of Business Research* 36(1): 15–23.

Myers, R. (1997). "Hidden agendas, power, and managerial assumptions in information systems development: an ethnographic study." *Information Technology & People* 10(3): 1997.

Neeley, A., Ed. (2002). *Business performance measurement: theory and practice*. Cambridge, U.K., Cambridge University Press.

Neumann (2000). "Risks in our information infrastructures: the tip of a titanic iceberg is still all that is visible." *Ubiquity: An ACM IT Magazine and Forum* 1(13).

Ngwenyama, O.K. and A.S. Lee (1997). "Communication richness in electronic mail: critical theory and the contextuality of meaning." *MIS Quarterly* 21(2): 145–167.

Nonaka, I. and H. Takeuchi (1995). *The knowledge-creating company: how Japanese companies create the dynamics of innovation*. New York, NY, Oxford University Press.

Norman, R. and R. Gregory (2003). "Paradoxes and pendulum swings: performance management in New Zealand's public sector." *Australian Journal of Public Administration* 62(4): 35–49.

Norrie, J. and D.H.T. Walker (2004). "A balanced scorecard approach to project management leadership." *Project Management Journal* 35 (4): 47–57.

O'Donnell, D., P. O'Regan, et al. (2003). "Human interaction: the critical source of intangible value." *Journal of Intellectual Capital* 4(1): 82–99.

Office of Government Commerce. (2004, September 2004). "Successful delivery toolkit." v4.5.2. Retrieved 13 January, 2005, from http://www.ogc.gov.uk/sdtoolkit/reference/deliverylifecycle/benefits_mgmt.html.

Olesen, K. and M.D. Myers (1999). "Trying to improve communication and collaboration with information technology: an action research project which failed." *Information Technology and People* 12(4): 317–332.

Oxford University Press. (2005). "AskOxford.com." Retrieved 29 January, 2005, from http://www.askoxford.com/?view=uk.

Patel, M.B. and P.W.G. Morris (1999). University of Manchester, UK, Centre for Research in the Management of Projects (CRMP).

Pellegrainelli, S. (1997). "Programme management: organizing project-based change." *International Journal of Project Management* 15(3): 141–150.

Pessemier, E.A. and N.R. Baker (1971). *Project and program decisions in research and development*. Lafayette, IN, Purdue University.

Peters, T.J. and R. Waterman (1982). *In search of excellence: lessons from America's best-run companies*. New York, NY, Harper and Row.

Pfeffer, J. and R.I. Sutton (2000). *The knowledge-doing gap: how smart companies turn knowledge into action*. Boston, MA, Harvard Business School Press.

Phelan, T.M. (2004). *The impact of effectiveness and efficiency on project success*. Hoboken, NJ, Stevens Institute of Technology.

Pinto, J.K. (1998). *Power & politics in project management*. Sylva, NC, Project Management Institute.

Pinto, J.K. and J.E. Prescott (1988). "Variations in critical success factors over the stages of the project life cycle." *Journal of Management* 14(1): 5–18.

Pinto, J.K. and D.P. Slevin (1988). "Project success: definitions and measurement techniques." *Project Management Journal* 19(1): 67–72.

PMI (1996). *A Guide to the Project Management Body of Knowledge*. Upper Darby, PA, Project Management Institute.

PMI (1996; 2000; 2004). *A guide to the project management body of knowledge*. Newtown Square, Philadelphia, PA, Project Management Institute.

PMI (2003). *Program/Portfolio Management Standard Project Charter*. Newton Square, PA, Project Management Institute.

PMI (2005). *The Standard for Portfolio Management—Second Edition Draft*. Newton Square, PA, Project Management Institute.

Porter, M.E. (1996). "What is strategy?" *Harvard Business Review* 74(6): 61–78.

Rapoport, R.N. (1970). "Three dilemmas in action research." *Human Relations* 23(4): 499–513.

Raz, T. (1993). "Introduction of the project management discipline in a software development organization." *IBM Systems Journal* 32(2): 265–277.

Reason, P. and H. Bradbury (2001*). Handbook of action research: participative inquiry & practice.* London, England, Sage.

Redmond, W. H. (1991). "When technologies compete: the role of externalities in nonlinear market response." *Journal of Product Innovation Management* 8(3): 170–183.

Remenyi, D. and M. Sherwood-Smith (1998). "Business benefits from information systems through an active benefits realisation programme." *International Journal of Project Management* 16(2): 81–98.

Robey, D. and M.L. Markus (1998). "Beyond rigor and relevance: producing consumable research about information systems." *Information Resources Management Journal* 11(1): 7–15.

Rosser, B. and K. Potter (2001). *IT Portfolio Management and Survey Results.* Stamford, CT, Gartner Group.

Rousell, P. (1991). *Third generation R&D: managing the link to corporate strategy.* Boston, MA, Harvard Business School.

Ruggles, R. (1998). "The State of Notion: Knowledge management in practice." *California Management Review* 40(3): 80–89.

Rumelt R.P., D.E. Schendel and D.J. Teece. (1994). *Fundamental issues in strategy.* Cambridge, MA, Harvard Business School Press: 9–47.

Rzasa, P.V., T.W. Faulkner, et al. (1990). "Analyzing R&D portfolios at Eastman Kodak." *Research Technology Management* 33(1): 27–32.

Saaty, T.L., P.C. Rogers, et al. (1980). "Portfolio selection through hierarchies." *The Journal of Portfolio Management* 6(3): 16–21.

Santos, B.L. (1989). *Selecting information systems projects: problems, solutions & challenges.* IEEE: System Sciences Conference, Hawaii.

Schon, D.A. (1983). *The reflective practitioner: how professionals think in action.* Aldershot, U.K., BasiAshgate ARENA Publishers.

Schwalbe, K. (2001). *Information Technology Project Management.* Boston, MA, Thompson: Course Technology.

Senge, P.M. (1990). *The fifth discipline: the art & practice of the learning organization.* New York, NY, Random House.

Shadish, William R. (1995). "The logic of generalization: five principles common to experiments and ethnographies." *American Journal of Community Psychology* 23(3): 419

Shank, M.E., A.C. Boynton, et al. (1985). "Critical success factor analysis as a methodology for MIS planning." *MIS Quarterly* 9(2): 121–129.

Sharpe, W.F. (1964). "Capital asset prices: a theory of market equilibrium under conditions of risk." *Journal of Finance* (19): 452–442.

Shead, B. (1998). "Outcomes and outputs: who's accountable for what?" *Accountability & Performance* 4(1): 89–100.

Shenhar, A.J. and D. Dvir (1996). "Chapter 32: long-term success dimensions in technology-based organizations." *Handbook of Technology Management*. New York, NY, McGraw Hill.

Shenhar, A.J., D. Dvir, et al. (2001). "Project success: a multidimensional strategic concept." *Long Range Planning: International Journal of Strategic Management* 34(6): 699–725.

Shenhar, A.J., O. Levy, et al. (1997). "Mapping the dimensions of project success." *Project Management Journal* 28(2): 5–13.

Simon, T. (2003). "What is benefit realization?" *The Public Manager* (Winter 2003-2004): 59–60.

Smith, M. (1998). "Innovation and the great IBM trade-off." *Management Accounting* 76(1): 24–26.

Snider, K.F. and M.E. Nissen (2003). "Beyond the body of knowledge: a knowledge-flow approach to project management theory and practice." *Project Management Journal* 34(2): 4–12.

Snowden, D. (2003). "Context, narrative and content: breaking away from the tacit and explicit words." Keynote address. Third International Conference on Culture and Change in Organizations, Penang, Malaysia.

Söderlund, J. (2004). "On the broadening scope of the research on projects: a review and a model for analysis." *International Journal of Project Management* 22(8): 655–667.

Souder, W.E. (1973). "Utility and perceived acceptability of R&D project selection methods." *Management Science* 19(12): 1384–1394.

Souder, W.E. (1975). "Achieving organizational consensus with respect to R&D selection criteria." *Management Science* 21(6): 669–681.

Souder, W.E. (1984). *Project selection and economic appraisal*. New York, NY, Van Nostrand Reinhold.

Steane, P.D. (1997). "Oils ain't oils! strategy across sectors." *International Journal of Public Sector Management* 10(6): 461–470.

Steiner, G.A. (1979). *Strategic planning: what every manager must know*. New York, NY, The Free Press.

Stewart, R.A. and S. Mohamed (2001). "Utilizing the balanced scorecard for IT/IS." *Construction Innovation* 1(3): 147–163.

Stewart, T.A. (1997). *Intellectual capital: the new wealth of organizations*. New York, NY, Doubleday.

Stewart, T.A. (1998). *Intellectual capital: the new wealth of organizations*. London, U.K., Nicholas Brealey Publishing.

Stewart, W.E. (2001). "Balanced scorecard for projects." *Project Management Journal* 32(1): 38–53.

Strauss, A. and J. Corbin (1998). *The basics of qualitative research: techniques and procedures for developing grounded theory*. Thousand Oaks, CA, Sage.

Susman, G.I. and R.D. Evered (1978). "An assessment of the scientific merits of action research." *Administrative Science Quarterly* 23(4): 582–603.

Sveiby, K.E. (1997). *The new organizational wealth: managing & measuring knowledge-based assets*. San Francisco, CA, Berrett-Koehler Publishers, Inc.

Szymcazk, C.C. and D.H.T. Walker (2003). "Boeing: a case study example of enterprise project management from a learning organization perspective." *The Learning Organization Journal* 10(3): 125–137.

Teece, D.J. (1998). "Capturing value from knowledge assets: the new economy, markets for know-how, and intangible assets." *California Management Review* 40(3): 55–80.

Thamhain, H.J. (1994). "Designing modern project management systems for a radically changing world." *Project Management Journal* 25(4): 6.

Thamhain, H.J. (2004). "Linkages of project environment to performance: lessons for team leadership." *International Journal of Project Management* 22(7): 533–544.

Thite, M. (1999). "Leadership styles in information technology projects." *International Journal of Project Management* 18(4): 235–241.

Thomas, H., T. Pollock, et al. (1999). "Global strategic analysis: frameworks and approaches." *The Academy of Management Executive* 13(1): 70–82.

Thomas, J., C. Delisle, and K. Jugdev (2002). *Selling Project Management to Senior Executives: Framing the Moves that Matter*. Newton Square, PA, Project Management Institute.

Trauth, E.M. (2001). "Choosing qualitative methods in IS research: lessons learned." *Qualitative research in IS: issues and trends*. E. M. Trauth. Hershey, PA, Idea Group Publishing: 271–287.

Tuman, J. (1986). "Success modeling: a technique for building a winning project team." *Project Management Institute's Annual Seminar & Symposium*, Montreal, PQ.

Turner, J.R. (1999). *The handbook of project-based management: improving the process for achieving strategic objectives*. Maidenhead, U.K., McGraw-Hill.

Turner, J.R. and R.A. Cochrane (1993). "The goals and methods matrix: coping with projects with ill-defined goals and/or methods of achieving them." *International Journal of Project Management* 11(2): 93–102.

Turner, J.R.; Grude, K.V. and L. Thurloway (1996). *The project manager as change agent: leadership, influence and negotiation*. London, U.K., McGraw-Hill.

Turner, J.R. and R. Muller (2005). "The project manager's leadership style as a success factor on projects: A literature review." *Project Management Journal* 36(1): 49–61

Ulri, B. and D. Ulri (2000). "Le management de projet et ses evolution en Amerique du Nord." *Revue Francaise de Gestion* (129): 21–31.

Van de Ven, A. and G.P. Huber (1990). "Longitudinal field research on change: theory and practice." *Organization Science* 1(3): 267–292.

Van Grembergen, Wim, S. De Haes and I. Amelinckx (2003). "Using CobIT and the Balanced Scorecard as Instruments for Service Level Management." *Information Systems Control Journal* 4(1).

Verwey, A. and D. Comninos (2002). "Business focused project management." *Management Services* 46(1): 14–22.

Walsham, G. (1993). "IS strategy and implementation: a case study of building society." *ACM SIGOIS Bulletin*. 12: 150–173.

Wang, S. (1997). "A synbook of natural language, semantic networks and business process modeling." *Canadian Journal of Administrative Sciences* 14(1): 79–92.

Ward, J. and R. Elvin (1999). "A new framework for managing IT-enabled business change." *Information Systems Journal* 9(3): 197–221.

Ward, J., P. Murray, et al. (2004). *Benefits management best practice guidelines*. Bedford, U.K., School of Management, Cranfield University.

Wateridge, J. (1998). "How can IS/IT projects be measured for success?" *International Journal of Project Management* 16(1): 59–63.

Wechsler, B. and R.W. Backoff (1987). "The dynamics of strategy in public organizations." *Journal of the American Planning Association* 53(1): 34–43.

Weick, K.E. (2001). *Making sense of the organization*. Oxford, U.K., Blackwell Publishers.

Weill, P. and M. Broadbent (1998). *Leveraging the new infrastructure*. Boston, MA, Harvard Business School Press.

Wells, W.G.J. (1998). "From the editor." *Project Management Journal* 29(2): 5.

West, D. and M.H. Stansfield (2001). "Structuring action and reflection in information systems action research studies using Checkland's FMA model." *Systemic Practice and Action Research* 14(3): 251–281.

Whitten, N. (1995). *Managing Software Development Projects: Formula for Success*, New York, NY, John Wiley & Sons.

Whittle, S. (2004). Make them walk your talk. *Computer Weekly*: 40–41.

Wideman, R.M. (1995). *Cost control of capital projects*. Richmond, BC, Canada, BiTech Publishers Ltd.

Wilemon, D. (2002). *Project management research: experiences and perspectives*. The Frontiers of Project Management. D.P. Slevin, D.I. Cleland and J.K. Pinto. Newtown Square, Project Management Institute: 57–71.

Wise, R.I. (1997). "The balanced scorecard approach to strategy management." *The Public Manager* (Fall 1997): 47–50.

Wood-Harper, A.T. (1985). "Research methods in information systems: using action research." *Research methods in information systems*. E. Mumford. Amsterdam, North-Holland: 161–191.

Yasin, M.M., J. Martin, et al. (2000). "An empirical investigation of international project management practices: the role of international experience." *Project Management Journal* 31(2): 20–30.

Yin, R.K. (2003). *Case study research: design & methods* (Third Edition). Thousand Oaks, CA, Sage.

Yukl, G. (1998). *Leadership in organizations*. Sydney, Australia, Prentice-Hall.

Zaleznik, A. (1977). "Managers and leaders: are they different?" *Harvard Business Review* 55(3): 67–78.

Zells, L. (1991). "Balancing trade-offs in quality, cost, schedule, resources, and risk." Project Management Institute Seminar/Symposium: paper presented September 28 to October 2.

Zeppou, M. and T. Sotirakou (2003). "The "STAIR" model: a comprehensive approach for managing and measuring government performance in the post-modern era." *The International Journal of Public Sector Management* 16(4): 320–332.

Zobel, A.M. and S.H. Wearne (1999). "Project management topic coverage in recent conferences." *Project Management Journal* 31(2): 32–37.

Zuber-Skerritt (2002). "A model for designing action learning and action research programs." *The Learning Organization Journal* 9(4): 143–149.

Index

Note: Page numbers appearing in **bold type** refer to table or figures.

The Best Resources in Leadership, HR, Management and Organizational Development